© 2016
Clement Ampadu
drampadu@hotmail.com

ISBN:978-1-365-56360-7
ID: 19695455
www.lulu.com

All rights reserved. No part of this publication may be produced or transmitted in any form or by any means, electronic or mechanical, including photocopying and recording, or in any information storage and retrieval system, without the prior written permission of the publisher.

Contents

	Dedication	3
1	**Higher-Order Hardy-Rogers Expanding Mapping Theorem in Partial Metric Spaces**	**4**
1.1	Brief Summary	4
1.2	Preliminaries	4
1.3	Main Results	8
1.4	Exercises	9
1.5	References	10
2	**Ψ-Higher-Order Contractions and Some Common r-Fixed Point Theorems in 0-Complete Partial Metric Spaces**	**11**
2.1	Brief Summary	11
2.2	Preliminaries	11
2.3	Main Results	15
2.4	Exercises	19
2.5	References	20
3	**r-Best-Proximity Point Theorems for Weak Generalization of Higher-Order Hardy-Rogers Type Mapping in Partial Metric Spaces**	**21**
3.1	Brief Summary	21
3.2	Preliminaries	21
3.3	Main Results	25
3.4	Exercises	28
3.5	References	29
4	**Graphic Weak Ψ-Higher Order Contractions with Application to Integral Equations**	**30**
4.1	Brief Summary	30
4.2	Preliminaries	30
4.3	Main Results	34
4.4	Exercises	42
4.5	References	43

Dedication

This book is dedicated to those who read it .

Clement Ampadu
November, 2016

Chapter 1

Higher-Order Hardy-Rogers Expanding Mapping Theorem in Partial Metric Spaces

1.1 Brief Summary

> **Abstract A.1 1**
>
> The notion of expanding mappings was introduced by Wang and colleagues [S. Z. Wang, B. Y. Li, Z. M. Gao, K. Iseki, Some fixed point theorems for expansion mappings, Math. Japonica. 29 (1984), 631-636] whom proved some fixed point theorems in complete metric spaces. On the other hand Daffer and Kaneko [P. Z. Daffer, H. Kaneko, On expansive mappings, Math. Japonica. 37 (1992), 733-735] defined an expanding condition for a pair of mappings and proved some common fixed point theorems for two mappings in complete metric spaces. In this paper we define expanding mappings in partial metric spaces analogous to expanding mappings in metric spaces, and establish some fixed point theorems concerning higher-order expanding maps.

1.2 Preliminaries

> **Definition A.1 1**
>
> Let X be a nonempty set. A function $\rho : X \times X \mapsto \mathbb{R}^+$ will be called a partial metric iff for all $x, y, z \in X$
>
> (a) $x = y \Leftrightarrow \rho(x,x) = \rho(x,y) = \rho(y,y)$
>
> (b) $\rho(x,x) \leq \rho(x,y)$
>
> (c) $\rho(x,y) = \rho(y,x)$
>
> (d) $\rho(x,z) \leq \rho(x,y) + \rho(y,z) - \rho(y,y)$
>
> Moreover, the pair (X, ρ) will be called a partial metric space

> **Remark A.2 1**
>
> If $\rho(x,y) = 0$, then from (a) and (b), $x = y$. However, if $x = y$, then $\rho(x,y)$ may not be zero

> **Example A.3 1**
>
> Let $\rho(x,y) = \max\{x,y\}$ for all $x,y \in \mathbb{R}^+$, then $(\mathbb{R}^+, \rho)$ is a partial metric space

> **Remark A.4 1**
>
> Examples of partial metric spaces from a computational point of view can be found in [S. Matthews, Partial metric topology, in: Proc. 8th Summer Conference on General Topology and Applications. Ann. New York Acad. Sci. 728 (1994), 183-197; S. Oltra, O. Valero, Banach's Fixed point theorem for partial metric spaces, Rend. Ist. Mat. Univ. Trieste. 36 (2004), 17-26; S. O'Neill, Partial metrics, valuations and domain theory, in: Proc. 11th Summer Conference on General Topology and Applications. Ann. New York Acad. Sci. 806 (1996), 304-315; O. Valero, On Banach fixed point theorems for partial metric spaces, Appl. Gen. Topol. 6 (2) (2005), 229-240]. Other generalization of partial metric spaces can be found in [S. O'Neill, Partial metrics, valuations and domain theory, in: Proc. 11th Summer Conference on General Topology and Applications. Ann. New York Acad. Sci. 806 (1996), 304-315; R. Heckmann, Approximation of metric spaces by partial metric spaces, Appl. Categ. Structures. 7 (1999), 71-83]

> **Remark A.5 1**
>
> Each partial metric ρ on X generates a T_0 topology $T(\rho)$ on X which has as a base the family of open ρ-balls $\{B_\rho(x;\epsilon) : x \in X; \epsilon > 0\}$, where, $B_p(x,\epsilon) = \{y \in X : \rho(x,y) < \rho(x,x) + \epsilon\}$ for all $x \in X$ and $\epsilon > 0$

From the remark immediately above, we have the following

> **Definition A.6 1**
>
> A sequence $\{x_n\}$ in a partial metric space (X,ρ) converges to a point $x \in X$ iff $\rho(x,x) = \lim_{n\to\infty} \rho(x,x_n)$

> **Definition A.7 1**
>
> A sequence $\{x_n\}$ in a partial metric space (X,ρ) is called a Cauchy sequence if $\lim_{n,m\to\infty} \rho(x_n, x_m) < \infty$

> **Definition A.8 1**
>
> A partial metric space (X,ρ) is said to be complete if every Cauchy sequence $\{x_n\}$ in X converges, with respect to $T(\rho)$, to a point $x \in X$ such that $\rho(x,x) = \lim_{n,m\to\infty} \rho(x_n, x_m)$

> **Remark A.9 1**
>
> Note that every closed subset of a complete partial metric space is complete

> **Definition A.10 1**
>
> Let (X,ρ) be a partial metric space. A function $f: X \mapsto X$ will be called r-continuous at a point $a \in X$, if for every sequence $\{x_n\} \in X$ which converges in (X,ρ) to a, the sequence $\{f^r x_n\}$ for any $r \in \mathbb{N}$ converges to $f^r a$, that is, for any $r \in \mathbb{N}$, $\rho(a,a) = \lim_{n\to\infty} \rho(x_n, a) \Rightarrow \rho(f^r a, f^r a) = \lim_{n\to\infty} \rho(f^r x_n, f^r a)$

> **Remark A.11 1**
>
> The r-continuity of the self-maps in partial metric spaces is in fact the sequential r-continuity as the above definition shows.

CHAPTER 1. HIGHER-ORDER HARDY-ROGERS EXPANDING MAPPING THEOREM IN PARTIAL METRIC SPACES

> **Remark A.12 1**
>
> If ρ is a partial metric on X, then the function $\rho^s : X \times X \mapsto \mathbb{R}^+$ given by $\rho^s(x,y) = 2\rho(x,y) - \rho(x,x) - \rho(y,y)$ is a metric on X. Moreover, (X, ρ^s) is a metric space.

> **Definition A.13 1**
>
> Let (X, ρ) be a partial metric space. A map $T : X \mapsto X$ is called an expanding mapping, if for every $x, y \in X$, there exists a number $k > 1$ such that $\rho(Tx, Ty) \geq k\rho(x,y)$

> **Definition A.14 1**
>
> Let (X, ρ) be a partial metric space and let the self-map $T : X \to X$ satisfy $\rho(T^r x, T^r y) \geq \sum_{q=0}^{r-1} c_q \rho(T^q x, T^q y)$ for all x, y in X and $r \in \mathbb{N}$. If $c_q > 1$ for all $0 \leq q \leq r-1$, then, $T : X \to X$ will be called an r-th order expanding mapping

The counterpart to Proposition 4.1 [Jeffery Ezearn, Higher-order Lipschitz Mappings, Fixed Point Theory and Applications (2015) 2015:88] , can be interpreted in the following way, for higher-order expanding mappings in the setting of partial metric spaces

> **Proposition A.15 1**
>
> Let (X, ρ) be a partial metric space, and let T be an rth-order expanding mapping on X. For every pair $x \neq y \in X$, define
>
> $$Z := Z(x,y) = \min_{0 \leq v \leq r-1} \beta^{-v} \frac{\rho(T^v x, T^v y)}{\rho(x,y)}$$
>
> then
>
> $$Z = \min_{n \in 0 \cup \mathbb{N}} \beta^{-n} \frac{\rho(T^n x, T^n y)}{\rho(x,y)}$$
>
> where $\beta > 1$

> **Remark A.16 1**
>
> Note that for all $n \in \mathbb{N}$, $\rho(T^n x, T^n y) \geq k\rho(T^{n-1} x, T^{n-1} y)$ is not expanding. By induction, we are lead to consider when $\rho(T^n x, T^n y) \geq k^n \rho(x,y)$ is expanding for all $n \in \mathbb{N}$ and $0 < k \leq 1$. If $x \neq y$, since, $0 < k \leq 1$, we observe that the sequence $b_n := k^{-n} \frac{\rho(T^n x, T^n y)}{\rho(x,y)}$ is bounded above by 1. Thus, if we denote by Z the minimum of this sequence, then, there exists $n \in \mathbb{N}$, call it r, such that, $M \leq b_r \leq 1$. Thus, the counterpart to Proposition 4.1[Jeffery Ezearn, Higher-order Lipschitz Mappings, Fixed Point Theory and Applications (2015) 2015:88] for higher-order expanding mappings is as above in the setting of partial metric spaces

Now we have the following alternate characterization of rth-order expanding mappings

> **Definition A.17 1**
>
> Let (X, ρ) be a partial metric space. A map $T : X \mapsto X$ will be called an rth-order expanding mapping on X, if $d(T^r x, T^r y) \geq Z\beta^r d(x,y)$ for all $x, y \in X$ and any $r \in \mathbb{N}$, where Z and β are given by Proposition A.15

> **Definition A.18 1**
>
> Two self-mappings f and g of a partial metric space (X, ρ) will be called r-commuting if $f^r g^r x = g^r f^r x$ for all $x \in X$ and any $r \in \mathbb{N}$

CHAPTER 1. HIGHER-ORDER HARDY-ROGERS EXPANDING MAPPING THEOREM IN PARTIAL METRIC SPACES

Definition A.19 1

Let $f, g : X \mapsto X$. If $w = f^r x = g^r x$ for some $x \in X$ and any $r \in \mathbb{N}$, then x will be called an r-coincidence point of f, g; w will be called a point of r-coincidence of f, g

Definition A.20 1

Let $f, g : X \mapsto X$. We will say that f, g are r-weakly compatible if they commute at their r-coincidence point, that is, $f^r x = g^r x$ for some $x \in X$ and any $r \in \mathbb{N}$, then, $f^r g^r x = g^r f^r x$

In the sequel we will need the following from [S. Matthews, Partial metric topology, in: Proc. 8th Summer Conference on General Topology and Applications. Ann. New York Acad. Sci. 728 (1994), 183-197; S. Oltra, O. Valero, Banach's fixed point theorem for partial metric spaces, Rend. Ist. Mat. Univ. Trieste. 36 (2004), 17-26]

Lemma A.21 1

Let (X, ρ) be a partial metric space.

(a) $\{x_n\}$ is a Cauchy sequence in (X, ρ) iff it is a Cauchy sequence in the metric space (X, ρ^s)

(b) A partial metric space (X, ρ) is complete iff the metric space (X, ρ^s) is complete. Furthermore, $\lim_{n \to \infty} \rho^s(a, x_n) = 0$ iff $\rho(a, a) = \lim_{n \to \infty} \rho(a, x_n) = \lim_{n,m \to \infty} \rho(x_n, x_m)$

From Definition A.1 [Ampadu, Clement(2016). Characterization Theorems Inspired by the Hardy-Rogers Map I: Some Results in Metric Spaces, lulu.com .ISBN: 1365101185, 9781365101182] we have the following

Definition A.22 1

Let (X, ρ) be a partial metric space. A map $T : X \mapsto X$ will be called an expanding Hardy-Rogers type map if it satisfies $\rho(Tx, Ty) \geq k[\rho(x, Tx) + \rho(y, Ty) + \rho(x, Ty) + \rho(y, Tx) + \rho(x, y)]$ for all $x, y \in X$ and $k > \frac{1}{5}$

Now the higher-order version of the above definition can be stated in the following way

Definition A.23 1

Let (X, ρ) be a partial metric space and let the self-map $T : X \to X$ satisfy $\rho(T^r x, T^r y) \geq \sum_{q=0}^{r-1} c_q[\rho(T^q x, T^{q+1} x) + \rho(T^q y, T^{q+1} y) + \rho(T^q x, T^{q+1} y) + \rho(T^q y, T^{q+1} x) + \rho(T^q x, T^q y)]$ for all x, y in X and $r \in \mathbb{N}$. If $c_q > \frac{1}{5}$ for all $0 \leq q \leq r-1$, then, $T : X \to X$ will be called an rth order expanding Hardy-Rogers type map.

Proposition A.24 1

Let (X, ρ) be a partial metric space, and let T be an rth-order expanding Hardy-Rogers type mapping on X. For every pair $x \neq y \in X$, define

$$Z := Z(x, y) = \min_{0 \leq v \leq r-1} \beta^{-v} \frac{\rho(T^v x, T^v y)}{\rho(x, Tx) + \rho(y, Ty) + \rho(x, Ty) + \rho(y, Tx) + \rho(x, y)}$$

then

$$Z = \min_{n \in 0 \cup \mathbb{N}} \beta^{-n} \frac{\rho(T^n x, T^n y)}{\rho(x, Tx) + \rho(y, Ty) + \rho(x, Ty) + \rho(y, Tx) + \rho(x, y)}$$

where $\beta > \frac{1}{5}$

> **Remark A.25 1**
>
> The counterpart to Proposition A.3[Ampadu, Clement(2016). Characterization Theorems Inspired by the Hardy-Rogers Map I: Some Results in Metric Spaces, lulu.com .ISBN: 1365101185, 9781365101182] for higher-order expanding Hardy-Rogers type map is as above in the setting of partial metric spaces

Now we have the following, which characterizes the expanding condition for higher-order Hardy-Rogers type map in an alternate way

> **Definition A.26 1**
>
> Let (X, ρ) be a partial metric space. A map $T : X \mapsto X$ will be called a higher-order expanding Hardy-Rogers type map if $\rho(T^r x, T^r y) \geq Z\beta^r[\rho(x, Tx) + \rho(y, Ty) + \rho(x, Ty) + \rho(y, Tx) + \rho(x, y)]$ for all $x, y \in X$, where Z and β are given by Proposition A.24

The following from [Xianjiu Huang, Chuanxi Zhu, Xi Wen, Fixed point theorems for expanding mappings in partial metric spaces, An. St. Univ. Ovidius Constanta, Vol. 20(1), 2012, 213-224] will be useful in the sequel

> **Lemma A.27 1**
>
> Let (X, ρ) be a partial metric space and $\{x_n\}$ be a sequence in X. If there exists $k \in (0, 1)$ such that $\rho(x_{n+1}, x_n) \leq k\rho(x_n, x_{n-1})$, $n = 1, 2, \cdots$, then $\{x_n\}$ is a Cauchy sequence in X

1.3 Main Results

> **Theorem A.1 1**
>
> Let (X, ρ) be a complete partial metric space and $T : X \mapsto X$ be an r-surjection, that is, T^r is surjective for any $r \in \mathbb{N}$. For any $r \in \mathbb{N}$, let Z and β be given by Proposition A.24 such that $\rho(T^r x, T^r y) \geq Z\beta^r[\rho(x, Tx) + \rho(y, Ty) + \rho(x, Ty) + \rho(y, Tx) + \rho(x, y)]$ for all $x, y \in X$ and $x \neq y$. Then T has an r-fixed point in X

> **Proof of Theorem A.1 1**
>
> Let $x_0 \in X$. Since T is an r-surjection, choose $x_1 \in X$ such that $T^r x_1 = x_0$, and in general choose $\{x_n\} \in X$ such that $x_{n-1} = T^r x_n$, $n = 1, 2, \cdots$. If there exist $n_0 \in \mathbb{N}$ such that $x_{n_0-1} = x_{n_0}$, then, x_{n_0} is an r-fixed point of T. So we assume that $x_{n-1} \neq x_n$ for all $n = 1, 2, \cdots$. Now observe that,
>
> $$\begin{aligned}\rho(x_{n-1}, x_n) &= \rho(T^r x_n, T^r x_{n+1}) \\ &\geq Z\beta^r [\rho(x_n, Tx_n) + \rho(x_{n+1}, Tx_{n+1}) + \rho(x_n, Tx_{n+1}) + \rho(x_{n+1}, Tx_n) \\ &\quad + \rho(x_n, x_{n+1})] \\ &= Z\beta^r [\rho(x_n, x_{n-1}) + \rho(x_{n+1}, x_n) + \rho(x_n, x_n) + \rho(x_{n+1}, x_{n-1}) + \rho(x_n, x_{n+1})] \\ &= Z\beta^r [\rho(x_n, x_{n-1}) + 2\rho(x_{n+1}, x_n) + \rho(x_{n+1}, x_{n-1}) + \rho(x_n, x_n)] \\ &\leq Z\beta^r [\rho(x_n, x_{n-1}) + 2\rho(x_{n+1}, x_n) + \rho(x_{n+1}, x_n) + \rho(x_n, x_{n-1}) - \rho(x_n, x_n) \\ &\quad + \rho(x_n, x_n)] \\ &= Z\beta^r [\rho(x_n, x_{n-1}) + 2\rho(x_{n+1}, x_n) + \rho(x_{n+1}, x_n) + \rho(x_n, x_{n-1})] \\ &= Z\beta^r [2\rho(x_n, x_{n-1}) + 3\rho(x_{n+1}, x_n)]\end{aligned}$$
>
> From the above one deduces that $\rho(x_{n+1}, x_n) \leq \gamma \rho(x_n, x_{n-1})$, where $\gamma := \frac{1-2Z\beta^r}{3Z\beta^r} < 1$. By Lemma A.27, $\{x_n\}$ is a Cauchy sequence in X. Since (X, ρ) is complete, then from Lemma A.21, (X, ρ^s) is complete, and so the sequence $\{x_n\}$ converges in the metric space (X, ρ^s), that is, there exists a point $z \in X$ such that $\lim_{n\to\infty} \rho^s(x_n, z) = 0$. Consequently, we can find $u \in X$ such that $z = T^r u$. Again from Lemma A.21, we have, $\rho(z, z) = \lim_{n\to\infty} \rho(x_n, z) = \lim_{n,m\to\infty} \rho(x_n, x_m)$. Moreover, since $\{x_n\}$ is Cauchy in the metric space (X, ρ^s), we have, $\lim_{n,m\to\infty} \rho^s(x_n, x_m) = 0$. On the other hand since $\max\{\rho(x_n, x_n), \rho(x_{n+1}, x_{n+1})\} \leq \rho(x_n, x_{n+1})$, then by induction with $\rho(x_{n+1}, x_n) \leq \gamma \rho(x_n, x_{n-1})$, we have, $\max\{\rho(x_n, x_n), \rho(x_{n+1}, x_{n+1})\} \leq \gamma^n \rho(x_0, x_1)$. It follows that
>
> $$\lim_{n\to\infty} \rho(x_n, x_n) = 0$$
>
> From the definition of ρ^s, we have, $\lim_{n,m\to\infty} \rho(x_n, x_m) = 0$. Now, it follows that $\rho(z, z) = \lim_{n\to\infty} \rho(x_n, z) = \lim_{n,m\to\infty} \rho(x_n, x_m) = 0$. Now we show that $u = z$. Observe that,
>
> $$\begin{aligned}\rho(x_n, z) &= \rho(T^r x_{n+1}, T^r u) \\ &\geq Z\beta^r [\rho(x_{n+1}, Tx_{n+1}) + \rho(u, Tu) + \rho(x_{n+1}, Tu) + \rho(u, Tx_{n+1}) + \rho(x_{n+1}, u)] \\ &= Z\beta^r [\rho(x_{n+1}, x_n) + \rho(u, Tu) + \rho(x_{n+1}, Tu) + \rho(u, x_n) + \rho(x_{n+1}, u)]\end{aligned}$$
>
> Taking limits in the above, we deduce that $0 = \rho(z, z) \geq Z\beta^r [3\rho(z, u)]$. Consequently, $\rho(u, z) = 0$, that is, $u = z = T^r u$

1.4 Exercises

> **Exercise A.1 1**
>
> Prove the following: Let (X, ρ) be a complete partial metric space, and $T : X \mapsto X$ be an r-surjection, that is, T^r is surjective for any $r \in \mathbb{N}$. Let Z and β be given by Proposition A.15 such that $\rho(T^r x, T^r y) \geq Z\beta^r \rho(x, y)$ for all $x, y \in X$. Then T has a unique r-fixed point in X, that is, there exists a unique $u \in X$ such that $T^r u = u$ for any $r \in \mathbb{N}$

> **Exercise A.2 1**
>
> Prove the following: Let (X, ρ) be a partial metric space. Let S and T be r-weakly compatible self-mappings of X and $T^r(X) \subseteq S^r(X)$ for any $r \in \mathbb{N}$. In Proposition A.15, let Z^* be a modification on Z and let β be as is given, such that, $\rho(S^r x, S^r y) \geq Z^* \beta^r \rho(Tx, Ty)$ for all $x, y \in X$. If one of the subspaces $T^r(X)$ or $S^r(X)$ is complete for any $r \in \mathbb{N}$, then S and T have a unique common r-fixed point in X

> **Exercise A.3 1**
>
> Deduce the following:
>
> (a) Exercise A.1 is the higher-order version of Corollary 2.1 [Xianjiu Huang, Chuanxi Zhu, Xi Wen, Fixed point theorems for expanding mappings in partial metric spaces, An. St. Univ. Ovidius Constanta, Vol. 20(1), 2012, 213-224]
>
> (b) Exercise A.2 is the higher-order version of Theorem 2.3 [Xianjiu Huang, Chuanxi Zhu, Xi Wen, Fixed point theorems for expanding mappings in partial metric spaces, An. St. Univ. Ovidius Constanta, Vol. 20(1), 2012, 213-224]

> **Exercise A.4 1**
>
> Using techniques of this chapter obtain the higher-order version of Theorem 2.1[Xianjiu Huang, Chuanxi Zhu, Xi Wen, Fixed point theorems for expanding mappings in partial metric spaces, An. St. Univ. Ovidius Constanta, Vol. 20(1), 2012, 213-224]

1.5 References

(1) S. Z. Wang, B. Y. Li, Z. M. Gao, K. Iseki, Some fixed point theorems for expansion mappings, Math. Japonica. 29 (1984), 631-636

(2) P. Z. Daffer, H. Kaneko, On expansive mappings, Math. Japonica. 37 (1992), 733-735

(3) S. Matthews, Partial metric topology, in: Proc. 8th Summer Conference on General Topology and Applications. Ann. New York Acad. Sci. 728 (1994), 183-197

(4) S. Oltra, O. Valero, Banach's Fixed point theorem for partial metric spaces, Rend. Ist. Mat. Univ. Trieste. 36 (2004), 17-26

(5) S. O'Neill, Partial metrics, valuations and domain theory, in: Proc. 11th Summer Conference on General Topology and Applications. Ann. New York Acad. Sci. 806 (1996), 304-315

(6) O. Valero, On Banach fixed point theorems for partial metric spaces, Appl. Gen. Topol. 6 (2) (2005), 229-240

(7) R. Heckmann, Approximation of metric spaces by partial metric spaces, Appl. Categ. Structures. 7 (1999), 71-83

(8) Jeffery Ezearn, Higher-order Lipschitz Mappings, Fixed Point Theory and Applications (2015) 2015:88

(9) Ampadu, Clement(2016). Characterization Theorems Inspired by the Hardy-Rogers Map I: Some Results in Metric Spaces, lulu.com . ISBN: 1365101185, 9781365101182

(10) Xianjiu Huang, Chuanxi Zhu, Xi Wen, Fixed point theorems for expanding mappings in partial metric spaces, An. St. Univ. Ovidius Constanta, Vol.20(1), 2012, 213-224

Chapter 2

Ψ-Higher-Order Contractions and Some Common r-Fixed Point Theorems in 0-Complete Partial Metric Spaces

2.1 Brief Summary

> **Abstract B.1 1**
>
> We obtain a theorem of common r-fixed point which under certain conditions generalizes the higher-order Hardy-Rogers type mapping Theorem, Theorem A.2 [Ampadu, Clement (2016). lulu.com, Characterization Theorems Inspired by the Hardy-Rogers Map I: Some Results in Metric Spaces. ISBN: 1365101185, 9781365101182] to partial metric spaces. Using a condition related to the higher-order Hardy Rogers type contraction, we establish a homotopy result in partial metric spaces.

2.2 Preliminaries

> **Definition B.1 1**
>
> Let X be a nonempty set. A function $\rho : X \times X \mapsto \mathbb{R}^+$ will be called a partial metric iff for all $x, y, z \in X$
>
> (a) $x = y \Leftrightarrow \rho(x,x) = \rho(x,y) = \rho(y,y)$
>
> (b) $\rho(x,x) \leq \rho(x,y)$
>
> (c) $\rho(x,y) = \rho(y,x)$
>
> (d) $\rho(x,z) \leq \rho(x,y) + \rho(y,z) - \rho(y,y)$
>
> Moreover, the pair (X, ρ) will be called a partial metric space

> **Remark B.2 1**
>
> If $\rho(x,y) = 0$, then from (a) and (b), $x = y$. However, if $x = y$, then $\rho(x,y)$ may not be zero

> **Example B.3 1**
>
> Let $\rho(x,y) = \max\{x,y\}$ for all $x, y \in \mathbb{R}^+$, then $(\mathbb{R}^+, \rho)$ is a partial metric space

Remark B.4 1

Each partial metric ρ on X generates a T_0 topology $T(\rho)$ on X which has as a base the family of open ρ-balls $\{B_\rho(x;\epsilon) : x \in X; \epsilon > 0\}$, where, $B_p(x,\epsilon) = \{y \in X : \rho(x,y) < \rho(x,x) + \epsilon\}$ for all $x \in X$ and $\epsilon > 0$

Remark B.5 1

If ρ is a partial metric on X, then the function $\rho^s : X \times X \mapsto \mathbb{R}^+$ given by $\rho^s(x,y) = 2\rho(x,y) - \rho(x,x) - \rho(y,y)$ is a metric on X. Moreover, (X, ρ^s) is a metric space.

From the remark immediately above, we have the following

Definition B.6 1

A sequence $\{x_n\}$ in a partial metric space (X, ρ) converges to a point $x \in X$ iff $\rho(x,x) = \lim_{n \to \infty} \rho(x, x_n)$

Definition B.7 1

A sequence $\{x_n\}$ in a partial metric space (X, ρ) is called a Cauchy sequence if $\lim_{n,m \to \infty} \rho(x_n, x_m)$ exists

Definition B.8 1

A partial metric space (X, ρ) is said to be complete if every Cauchy sequence $\{x_n\}$ in X converges, with respect to $T(\rho)$, to a point $x \in X$ such that $\rho(x,x) = \lim_{n,m \to \infty} \rho(x_n, x_m)$

Definition B.9 1

A sequence $\{x_n\}$ in a partial metric space (X, ρ) is called 0-Cauchy if $\lim_{n,m \to \infty} \rho(x_n, x_m) = 0$

Definition B.10 1

We say that (X, ρ) is 0-Complete if every 0-Cauchy sequence in X converges, with respect to $T(\rho)$, to a point $x \in X$ such that $\rho(x,x) = 0$

Remark B.11 1

Every closed subset of a 0-complete partial metric space is 0-complete

Example B.12 1

Let $\mathbb{Q}$ denote the set of rational numbers, then, the partial metric space $(\mathbb{Q} \cap [0, \infty), \max\{x, y\})$ is 0-complete but not complete

In the sequel we will need the following from [S. Matthews, Partial metric topology, in: Proc. 8th Summer Conference on General Topology and Applications. Ann. New York Acad. Sci. 728 (1994), 183-197; S. Oltra, O. Valero, Banach's fixed point theorem for partial metric spaces, Rend. Ist. Mat. Univ. Trieste. 36 (2004), 17-26]

Lemma B.13 1

Let (X, ρ) be a partial metric space.

(a) $\{x_n\}$ is a Cauchy sequence in (X, ρ) iff it is a Cauchy sequence in the metric space (X, ρ^s)

(b) A partial metric space (X, ρ) is complete iff the metric space (X, ρ^s) is complete. Furthermore, $\lim_{n \to \infty} \rho^s(a, x_n) = 0$ iff $\rho(a,a) = \lim_{n \to \infty} \rho(a, x_n) = \lim_{n,m \to \infty} \rho(x_n, x_m)$

In the sequel we will need the following from [Cristina Di Bari and Pasquale Vetro, Common Fixed Points for ψ-contractions on Partial Metric Spaces, Hacettepe Journal of Mathematics and Statistics Volume 42 (6) (2013), 591-598]

Lemma B.14 1

Let (X, ρ) be a partial metric space and $\{x_n\} \subset X$. If $x_n \to x \in X$ and $\rho(x,x) = 0$, then, $\lim_{n \to \infty} \rho(x_n, z) = \rho(x, z)$ for all $z \in X$

Remark B.15 1

Define $\rho(x, A) = \inf\{\rho(x, a) : a \in A\}$. Then $a \in \bar{A}$ iff $\rho(a, A) = \rho(a, a)$, where $\bar{A}$ denotes the closure of A. From $\rho^s(x, a) = 2\rho(x, a) - \rho(x, x) - \rho(a, a) \leq 2\rho(x, a)$ for every $a \in A$, we deduce that $\rho^s(x, A) \leq 2\rho(x, A)$

Definition B.16 1

Let X be a nonempty set and $T, f : X \mapsto X$. For any $r \in \mathbb{N}$, we will say

(a) T, f are r-weakly compatible if they commute at their r-coincidence point, that is, $T^r f^r x = f^r T^r x$ whenever $T^r x = f^r x$

(b) $y \in X$ is a point of r-coincidence of T, f if there exists a point $x \in X$ such that $y = T^r x = f^r x$

Definition B.17 1

Let (X, ρ) be a partial metric space and $T, f : X \mapsto X$ be such that $T^r X \subset f^r X$ for any $r \in \mathbb{N}$. For every $x_0 \in X$, let $\{x_n\} \subset X$ be defined by $f^r x_n = T^r x_{n-1}$ for all $n \in \mathbb{N}$ and any $r \in \mathbb{N}$, then we will say $\{T^r x_n\}$ is a r-T-f sequence of initial point x_0

In the sequel we will need the following from [Cristina Di Bari and Pasquale Vetro, Common Fixed Points for ψ-contractions on Partial Metric Spaces, Hacettepe Journal of Mathematics and Statistics Volume 42 (6) (2013), 591-598]

Lemma B.18 1

For every function $\psi : [0, \infty) \mapsto [0, \infty)$, let ψ^n be the nth iterate of ψ. Then the following holds: if ψ is nondecreasing, then for each $t > 0$, $\lim_{n \to \infty} \psi^n(t) = 0$ implies $\psi(t) < t$

Definition B.19 1

Let (X, ρ) be a partial metric space, and $T, f : X \mapsto X$. We will say that T is an rth order contraction with respect to f if

$$\rho(T^r x, T^r y) \leq \sum_{q=0}^{r-1} c_q \max\{\rho(f^{q+1}x, f^{q+1}y), \rho(f^{q+1}x, T^{q+1}x), \rho(f^{q+1}y, T^{q+1}y),$$
$$\frac{1}{2}[\rho(f^{q+1}x, T^{q+1}y) + \rho(f^{q+1}y, T^{q+1}x)]\}$$

for all $x, y \in X$, $0 \leq c_q < 1$ and $0 \leq q \leq r-1$

Proposition B.20 1

Let (X, ρ) be a partial metric space, and let T be an rth-order contraction with respect to f on X. For every pair $x \neq y \in X$, define

$$Z := Z(x,y) = \max_{0 \leq v \leq r-1} \beta^{-v} \frac{\rho(T^v x, T^v y)}{\max\{\rho(fx, fy), \rho(fx, Tx), \rho(fy, Ty), \frac{1}{2}[\rho(fx, Ty) + \rho(fy, Tx)]\}}$$

then

$$Z = \max_{n \in 0 \cup \mathbb{N}} \beta^{-n} \frac{\rho(T^n x, T^n y)}{\max\{\rho(fx, fy), \rho(fx, Tx), \rho(fy, Ty), \frac{1}{2}[\rho(fx, Ty) + \rho(fy, Tx)]\}}$$

where $\beta \in [0, 1)$

Definition B.21 1

Let (X, ρ) be a partial metric space, and $T, f : X \mapsto X$. We will say that T is an rth-order Hardy-Rogers type contraction with respect to f if

$$\rho(T^r x, T^r y) \leq \sum_{q=0}^{r-1} c_q [\rho(f^{q+1} x, f^{q+1} y) + \rho(f^{q+1} x, T^{q+1} x) + \rho(f^{q+1} y, T^{q+1} y) \\ + \rho(f^{q+1} x, T^{q+1} y) + \rho(f^{q+1} y, T^{q+1} x)]$$

for all $x, y \in X$, $0 \leq c_q < \frac{1}{5}$ and $0 \leq q \leq r-1$

Proposition B.22 1

Let (X, ρ) be a partial metric space, and let T be an rth-order Hardy-Rogers type contraction with respect to f on X. For every pair $x \neq y \in X$, define

$$Q := Q(x,y) = \max_{0 \leq v \leq r-1} \zeta^{-v} \frac{\rho(T^v x, T^v y)}{\rho(fx, fy) + \rho(fx, Tx) + \rho(fy, Ty) + \rho(fx, Ty) + \rho(fy, Tx)}$$

then

$$Q = \max_{n \in 0 \cup \mathbb{N}} \zeta^{-n} \frac{\rho(T^n x, T^n y)}{\rho(fx, fy) + \rho(fx, Tx) + \rho(fy, Ty) + \rho(fx, Ty) + \rho(fy, Tx)}$$

where $\zeta \in [0, \frac{1}{5})$

Remark B.23 1

If f is the identity in Definition B.21, then we obtain Definition A.2 [Ampadu, Clement (2016). lulu.com . Characterization Theorems Inspired by the Hardy-Rogers Map I: Some Results in Metric Spaces. ISBN: 1365101185, 9781365101182] in the setting of partial metric spaces. Similarly, if f is the identity in Proposition B.22, then we obtain Proposition A.3 [Ampadu, Clement (2016). lulu.com . Characterization Theorems Inspired by the Hardy-Rogers Map I: Some Results in Metric Spaces. ISBN: 1365101185, 9781365101182] in the setting of partial metric spaces.

2.3 Main Results

Lemma B.1 1

Let X be a nonempty set and the mappings $T, f : X \mapsto X$ have a unique point of r-coincidence v in X. If T and f are r-weakly compatible, then T and f have a unique common r-fixed point

Proof of Lemma B.1 1

Since v is a point of r-coincidence of T and f, it follows that $v = f^r u = T^r u$ for some $u \in X$ and any $r \in \mathbb{N}$. By r-weak compatibility of T and f, we have, $T^r v = T^r f^r u = f^r T^r u = f^r v$. It follows that $T^r v = f^r v = w(say)$. Thus w is a point of r-coincidence of T and f. Therefore $v = w$ by uniqueness. Thus, v is a unique common r-fixed point of T and f.

Theorem B.2 1

Let (X, ρ) be a partial metric space, and $T, f : X \mapsto X$ with $T^r(X) \subset f^r(X)$ for any $r \in \mathbb{N}$. Assume that

$$\rho(T^r x, T^r y) \leq \psi(\max\{\rho(fx, fy), \rho(fx, Tx), \rho(fy, Ty), \frac{1}{2}[\rho(fx, Ty) + \rho(fy, Tx)]\})$$

for all $x, y \in X$ and any $r \in \mathbb{N}$, where $\psi : [0, \infty) \mapsto [0, \infty)$ is right continuous, nondecreasing function such that $\sum_{n=1}^{\infty} \psi^n(t) < \infty$ for all $t > 0$. If $T^r(X)$ or $f^r(X)$ is a 0-complete subspace of X for any $r \in \mathbb{N}$, then T and f have a unique point of r-coincidence. Moreover, if T and f are r-weakly compatible, then T and f have a unique r-fixed point.

Proof of Theorem B.2 1

Fix $x_0 \in X$. We prove that the r-T-f sequence $\{T^r x_n\}$ of initial point x_0 is a Cauchy sequence in $T^r(X)$ for any $r \in \mathbb{N}$. If $T^r x_n = T^r x_{n-1}$ for some $n \in \mathbb{N}$ and any $r \in \mathbb{N}$, then $T^r x_n = T^r x_m$ for all $m \in \mathbb{N}$ with $m > n$ so $\{T^r x_n\}$ is a Cauchy sequence. Suppose that $T^r x_n \neq T^r x_{n-1}$ for all $n \in \mathbb{N}$ and any $r \in \mathbb{N}$. Observe that

$$\rho(T^r x_{n+1}, T^r x_n) \leq \psi(\max\{\rho(fx_{n+1}, fx_n), \rho(fx_{n+1}, Tx_{n+1}), \rho(fx_n, Tx_n),$$
$$\frac{1}{2}[\rho(fx_{n+1}, Tx_n) + \rho(fx_n, Tx_{n+1})]\})$$
$$= \psi(\max\{\rho(Tx_{n-1}, Tx_n), \rho(Tx_n, Tx_{n+1}), \frac{1}{2}[\rho(Tx_n, Tx_n)$$
$$+ \rho(Tx_{n-1}, Tx_{n+1})]\})$$

However, from (d) of Definition B.1, we have, $\rho(Tx_n, Tx_n) + \rho(Tx_{n-1}, Tx_{n+1}) \leq \rho(Tx_{n-1}, Tx_n) + \rho(Tx_n, Tx_{n+1})$, and thus, $\rho(T^r x_{n+1}, T^r x_n) \leq \psi(\max\{\rho(Tx_{n-1}, Tx_n), \rho(Tx_n, Tx_{n+1})\})$. Now if $\max\{\rho(Tx_{n-1}, Tx_n), \rho(Tx_n, Tx_{n+1})\} = \rho(Tx_n, Tx_{n+1}) \leq \rho(T^r x_n, T^r x_{n+1})$, then we get a contradiction. Thus, it follows that $\rho(T^r x_{n+1}, T^r x_n) \leq \psi(\rho(Tx_{n-1}, Tx_n))$, and by induction we have, $\rho(T^r x_{n+1}, T^r x_n) \leq \psi^n(\rho(Tx_0, Tx_1)) \leq \psi^n(\rho(T^r x_0, T^r x_1))$ for all $n \in \mathbb{N}$ and any $r \in \mathbb{N}$. Fix $\epsilon > 0$ and choose $n(\epsilon) \in \mathbb{N}$ such that $\sum_{n \geq n(\epsilon)} \psi^n(\rho(T^r x_1, T^r x_0)) < \epsilon$. Now, for all $n \geq n(\epsilon)$ and all $k \in \mathbb{N}$, we have,

$$\rho(T^r x_{n+k}, T^r x_n) \leq \sum_{j=1}^{k} \rho(T^r x_{n+j}, T^r x_{n+j-1}) - \sum_{j=1}^{k-1} \rho(T^r x_{n+j}, T^r x_{n+j})$$
$$\leq \sum_{j=1}^{k} \rho(T^r x_{n+j}, T^r x_{n+j-1})$$
$$\leq \sum_{n \geq n(\epsilon)} \psi^n(\rho(T^r x_1, T^r x_0))$$
$$\leq \epsilon$$

It follows that $\lim_{n,m \to \infty} \rho(T^r x_n, T^r x_m) = 0$ and hence $\{T^r x_n\}$ is a 0-Cauchy sequence in the partial metric space (X, ρ). Suppose that $T^r X$ is a 0-complete subspace of (X, ρ), then there exists $y \in T^r X \subset f^r X$ such that $\rho(y, y) = \lim_{n \to \infty} \rho(T^r x_n, y) = \lim_{n \to \infty} \rho(f^r x_n, y) = \lim_{n,m \to \infty} \rho(T^r x_n, T^r x_m) = 0$. Note that this also holds if $f^r X$ is a 0-complete subspace of (X, ρ) with $y \in f^r X$. Let $u \in X$ be such that $y = f^r u$. We show that y is a point of r-coincidence of T and f. If not, then $\rho(f^r u, T^r u) > 0$, on the other hand, $\rho(f^r u, T^r u) \leq \max\{\rho(fx_n, fu), \rho(fx_n, Tx_n), \rho(fu, Tu), \frac{1}{2}[\rho(fx_n, Tu) + \rho(fu, Tx_n)]\}$ for all $n \in \mathbb{N}$. Thus from Lemma B.14, and the right continuity of ψ, if we take limits as $n \to \infty$ in the inequality below,

$$\rho(T^r x_n, T^r u) \leq \psi(\max\{\rho(fx_n, fu), \rho(fx_n, Tx_n), \rho(fu, Tu), \frac{1}{2}[\rho(fx_n, Tu) + \rho(fu, Tx_n)]\})$$

we deduce that $\rho(f^r u, T^r u) \leq \psi(\rho(f^r u, T^r u)) < \rho(f^r u, T^r u)$, which is a contradiction, thus, $\rho(f^r u, T^r u) = 0$, that is, $f^r u = T^r u$. Hence, it follows that $y = f^r u = T^r u$ is a point of r-coincidence of T and f. If $z \in X$, with $z = f^r s = T^r s$, is another point of r-coincidence of T, f, then $z = y$. Suppose not, then we observe that

$$\rho(T^r u, T^r s) \leq \psi(\max\{\rho(fs, fu), \rho(fu, Tu), \rho(fs, Ts), \frac{1}{2}[\rho(fu, Ts) + \rho(fs, Tu)]\})$$
$$= \psi(\rho(Tu, Tus)) \leq \rho(Tu, Ts) \leq \rho(T^r u, T^r s)$$

which is a contradiction, thus, $y = z$, and y is a unique point of r-coincidence of T, f. By Lemma B.1, we deduce that y is a unique common r-fixed point of T and f

If in the previous theorem, we take $\psi(t) = Z\beta^r t$ where Z and β are given by Proposition B.20, then we obtain the following

> **Theorem B.3 1**
>
> Let (X, ρ) be a partial metric space and $T, f : X \mapsto X$ with $T^r X \subset f^r X$ for any $r \in \mathbb{N}$. Assume that
>
> $$\rho(T^r x, T^r y) \leq Z\beta^r \max\{\rho(fx, fy), \rho(fx, Tx), \rho(fy, Ty), \frac{1}{2}[\rho(fx, Ty) + \rho(fy, Tx)]\}$$
>
> for all $x, y \in X$, where Z and β are given by Proposition B.20. If $T^r X$ or $f^r X$ is a 0-complete subspace of X, then T and f have a unique point of r-coincidence. Moreover, if T and f are r-weakly compatible, then T and f have a unique r-fixed point

Since

$$Z\beta^r \max\{\rho(fx, fy), \rho(fx, Tx), \rho(fy, Ty), \frac{1}{2}[\rho(fx, Ty) + \rho(fy, Tx)]\}$$
$$\leq Q\zeta^r [\rho(fx, fy) + \rho(fx, Tx) + \rho(fy, Ty) + \rho(fx, Ty) + \rho(fy, Tx)]$$

then we have the following

> **Theorem B.4 1**
>
> Let (X, ρ) be a partial metric space and $T, f : X \mapsto X$ with $T^r X \subset f^r X$ for any $r \in \mathbb{N}$. Assume that
>
> $$\rho(T^r x, T^r y) \leq Q\zeta^r [\rho(fx, fy) + \rho(fx, Tx) + \rho(fy, Ty) + \rho(fx, Ty) + \rho(fy, Tx)]$$
>
> for all $x, y \in X$, where Q and ζ are given by Proposition B.22. If $T^r X$ or $f^r X$ is a 0-complete subspace of X, then T and f have a unique point of r-coincidence. Moreover, if T and f are r-weakly compatible, then T and f have a unique r-fixed point

> **Remark B.5 1**
>
> Note that the above theorem generalizes the higher-order Hardy-Rogers Mapping Theorem, Theorem A.2 [Ampadu, Clement (2016). lulu.com . Characterization Theorems Inspired by the Hardy-Rogers Map I: Some Results in Metric Spaces. ISBN: 1365101185, 9781365101182] to the setting of partial metric spaces

Finally the homotopy result is given as follows

> **Theorem B.6 1**
>
> Let (X, ρ) be a 0-complete partial metric space, U an open subset of X. Let $H : \bar{U} \times [0, 1] \mapsto X$ be such that
>
> (a) $x \neq H^r(x, \lambda)$ for all $x \in \partial U$, $\lambda \in [0, 1]$, and any $r \in \mathbb{N}$, where ∂U is the boundary of U in X
>
> (b) there exists Q and ζ given by Proposition B.22, a continuous function $\xi : [0, 1] \mapsto \mathbb{R}$, and any $r \in \mathbb{N}$ such that $\rho(H^r(x, \lambda), H^r(y, \mu)) \leq Q\zeta^r [\rho(x, y) + \rho(x, H(x, \lambda)) + \rho(y, H(y, \mu)) + \rho(x, H(y, \mu)) + \rho(y, H(x, \lambda))] + |\xi(\lambda) - \xi(\mu)|$ for all $x, y \in \bar{U}$ and $\lambda, \mu \in [0, 1]$
>
> If $H(\cdot, 0)$ has a r-fixed point in U, then $H(\cdot, \lambda)$ has a r-fixed point in U for every $\lambda \in [0, 1]$

CHAPTER 2. Ψ-HIGHER-ORDER CONTRACTIONS AND SOME COMMON R-FIXED POINT THEOREMS IN 0-COMPLETE PARTIAL METRIC SPACES

Proof of Theorem B.6 1

Consider the set $\Lambda = \{\lambda \in [0,1] : x = H^r(x,\lambda) \text{ for some } x \in U \text{ and any } r \in \mathbb{N}\}$. We note that $0 \in \Lambda$, since $H(\cdot, 0)$ has a r-fixed point in U. We will show that Λ is both closed and open in $[0,1]$ and hence by connectedness we have that $\Lambda = [0,1]$. We show that Λ is closed in $[0,1]$. Let $\{\lambda_n\} \subset \Lambda$ with $\lambda_n \to \lambda \in [0,1]$ as $n \to \infty$. Now for each $n \in \mathbb{N}$ and any $r \in \mathbb{N}$, there is $x_n \in U$ such that $x_n = H^r(x_n, \lambda_n)$. Now,

$$\rho(x_n, x_m) = \rho(H^r(x_n, \lambda_n), H^r(x_m, \lambda_m))$$
$$\leq Q\zeta^r[\rho(x_n, x_m) + \rho(x_n, H(x_n, \lambda_n)) + \rho(x_m, H(x_m, \lambda_m)) + \rho(x_n, H(x_m, \lambda_m))$$
$$+ \rho(x_m, H(x_n, \lambda_n))] + |\xi(\lambda_n) - \xi(\lambda_m)|$$
$$\leq Q\zeta^r[\rho(x_n, x_m) + \rho(x_n, H(x_n, \lambda_n)) + \rho(x_m, H(x_m, \lambda_m)) + \rho(x_n, x_m)$$
$$+ \rho(x_m, H(x_m, \lambda_m))$$
$$- \rho(x_m, x_m) + \rho(x_m, x_n) + \rho(x_n, H(x_n, \lambda_n)) - \rho(x_n, x_n)] + |\xi(\lambda_n) - \xi(\lambda_m)|$$
$$= Q\zeta^r[\rho(x_n, x_m) + \rho(x_n, x_n) + \rho(x_m, x_m) + \rho(x_n, x_m) + \rho(x_m, x_m)$$
$$- \rho(x_m, x_m) + \rho(x_m, x_n) + \rho(x_n, x_n) - \rho(x_n, x_n)] + |\xi(\lambda_n) - \xi(\lambda_m)|$$
$$= Q\zeta^r[\rho(x_n, x_m) + \rho(x_n, x_n) + \rho(x_m, x_m) + \rho(x_n, x_m) + \rho(x_m, x_n)]$$
$$+ |\xi(\lambda_n) - \xi(\lambda_m)|$$
$$\leq 5Q\zeta^r \rho(x_n, x_m) + |\xi(\lambda_n) - \xi(\lambda_m)|$$

From the above, we have, $\rho(x_n, x_m) \leq \frac{1}{1-5Q\zeta^r}|\xi(\lambda_n) - \xi(\lambda_m)|$, and since $\lim_{n,m\to\infty} \rho(x_n, x_m) = 0$, it follows that $\{x_n\}$ is a 0-Cauchy sequence. As X is 0-complete, there exists $x \in \bar{U}$ with $\rho(x,x) = \lim_{n\to\infty} \rho(x_n, x) = \lim_{n,m\to\infty} \rho(x_n, x_m) = 0$. Now observe that,

$$\rho(x_n, H^r(x, \lambda)) = \rho(H^r(x_n, \lambda_n), H^r(x, \lambda))$$
$$\leq Q\zeta^r[\rho(x_n, x) + \rho(x_n, H(x_n, \lambda_n)) + \rho(x, H(x, \lambda)) + \rho(x_n, H(x, \lambda))$$
$$+ \rho(x, H(x_n, \lambda_n))] + |\xi(\lambda_n) - \xi(\lambda)|$$
$$= Q\zeta^r[\rho(x_n, x) + \rho(x_n, x_n) + \rho(x, H(x, \lambda)) + \rho(x_n, H(x, \lambda))$$
$$+ \rho(x, x_n)] + |\xi(\lambda_n) - \xi(\lambda)|$$
$$\leq Q\zeta^r[\rho(x_n, x) + \rho(x_n, x_n) + \rho(x, x_n) + \rho(x_n, H(x, \lambda)) - \rho(x_n, x_n)$$
$$+ \rho(x_n, H(x, \lambda))$$
$$+ \rho(x, x_n)] + |\xi(\lambda_n) - \xi(\lambda)|$$
$$\leq Q\zeta^r[3\rho(x_n, x) + 2\rho(x_n, H^r(x, \lambda))] + |\xi(\lambda_n) - \xi(\lambda)|$$

From the above we deduce that

$$\rho(x_n, H^r(x, \lambda)) \leq \frac{3Q\zeta^r}{1 - 2Q\zeta^r}\rho(x_n, x) + \frac{1}{1-2Q\zeta^r}|\xi(\lambda_n) - \xi(\lambda)|$$

and taking limits in the above inequality we obtain $\rho(x, H^r(x, \lambda)) = 0$, that is, $x = H^r(x, \lambda)$, and so $\lambda \in \Lambda$, that is, Λ is closed in $[0, 1]$. Now we show that Λ is open in $[0, 1]$. Let $\lambda_0 \in \Lambda$ and $x_0 \in U$ with $x_0 = H^r(x_0, \lambda_0)$ for any $r \in \mathbb{N}$. Let $s = \rho(x_0, \partial U) = \inf\{\rho(x_0, x) : x \in \partial U\} > \rho(x_0, x_0)$. Fix $\epsilon \in (0, (1 - 5Q\zeta^r)s)$ and let $p > 0$ be such that for each $\lambda \in (\lambda_0 - p, \lambda_0 + p)$, we have, $|\xi(\lambda) - \xi(\lambda_0)| < \epsilon$. Now for fixed $\lambda \in (\lambda_0 - p, \lambda_0 + p)$ and $x \in Y = \{x \in X : \rho(x, x_0) \leq s\}$, we have,

> **Proof of Theorem B.6 continued 1**
>
> $$\begin{aligned}\rho(x_0, H^r(x,\lambda)) &= \rho(H^r(x_0,\lambda_0), H^r(x,\lambda)) \\ &\leq Q\zeta^r[\rho(x_0,x) + \rho(x_0, H(x_0,\lambda_0)) + \rho(x, H(x,\lambda)) + \rho(x_0, H(x,\lambda)) \\ &\quad + \rho(x, H(x_0,\lambda_0))] + |\xi(\lambda_0) - \xi(\lambda)| \\ &= Q\zeta^r[\rho(x_0,x) + \rho(x_0,x_0) + \rho(x, H(x,\lambda)) + \rho(x_0, H(x,\lambda)) \\ &\quad + \rho(x,x_0)] + |\xi(\lambda_0) - \xi(\lambda)| \\ &\leq Q\zeta^r[\rho(x_0,x) + \rho(x_0,x_0) + \rho(x,x_0) + \rho(x_0, H(x,\lambda)) - \rho(x_0,x_0) \\ &\quad + \rho(x_0, H(x,\lambda)) + \rho(x,x_0)] + |\xi(\lambda_0) - \xi(\lambda)| \\ &\leq Q\zeta^r[3\rho(x_0,x) + 2\rho(x_0, H^r(x,\lambda))] + |\xi(\lambda_0) - \xi(\lambda)|\end{aligned}$$
>
> Consequently, we have,
>
> $$\rho(x_0, H^r(x,\lambda)) \leq \frac{3Q\zeta^r}{1 - 2Q\zeta^r}\rho(x_0,x) + \frac{1}{1 - 2Q\zeta^r}|\xi(\lambda_0) - \xi(\lambda)| \leq s$$
>
> Thus for each fixed $\lambda \in (\lambda_0 - p, \lambda_0 + p)$ the mapping $H(\cdot, \lambda) : Y \mapsto Y$ has a r-fixed point in $\bar{U}$. However, this r-fixed point must be in U by condition (a). This implies that Λ is open in $[0,1]$. Thus, $\Lambda = [0,1]$ and $H(\cdot, \lambda)$ has a r-fixed point in U for every $\lambda \in [0,1]$

2.4 Exercises

> **Exercise B.1 1**
>
> Prove the following: Let (X, ρ) be a partial metric space and T, f be mappings on X with $T^r X \subset f^r X$ for any $r \in \mathbb{N}$. Assume there exists an effective bound $W \geq 1$ and $s \in [0,1)$ such that $\rho(T^r x, T^r y) \leq W s^r \rho(fx, fy)$ for all $x, y \in X$. If $T^r X$ or $f^r X$ is a 0-complete subspace of X for any $r \in \mathbb{N}$, then T and f have a unique point of r-coincidence. Moreover if T and f are r-weakly compatible then T and f have a unique r-fixed point.

> **Exercise B.2 1**
>
> Prove the following: Let (X, ρ) be a partial metric space and T, f be mappings on X with $T^r X \subset f^r X$ for any $r \in \mathbb{N}$. Assume there exists an effective bound $R \geq 1$ and $q \in [0,1)$ such that $\rho(T^r x, T^r y) \leq R q^r \max\{\rho(fx, Tx), \rho(fy, Ty)\}$ for all $x, y \in X$. If $T^r X$ or $f^r X$ is a 0-complete subspace of X for any $r \in \mathbb{N}$, then T and f have a unique point of r-coincidence. Moreover if T and f are r-weakly compatible then T and f have a unique r-fixed point.

> **Exercise B.3 1**
>
> Prove the following: Let (X, ρ) be a partial metric space and T, f be mappings on X with $T^r X \subset f^r X$ for any $r \in \mathbb{N}$. Assume there exists an effective bound $J \geq 1$ and $z \in [0, \frac{1}{3})$ such that $\rho(T^r x, T^r y) \leq J z^r [\rho(fx, Tx) + \rho(fy, Ty) + \rho(fx, fy)]$ for all $x, y \in X$. If $T^r X$ or $f^r X$ is a 0-complete subspace of X for any $r \in \mathbb{N}$, then T and f have a unique point of r-coincidence. Moreover if T and f are r-weakly compatible then T and f have a unique r-fixed point.

> **Exercise B.4 1**
>
> Let (X, ρ) be a 0-complete partial metric space, U an open subset of X. Let $H : \bar{U} \times [0,1] \mapsto X$ be such that
>
> (a) $x \neq H^r(x, \lambda)$ for all $x \in \partial U$, $\lambda \in [0,1]$, and any $r \in \mathbb{N}$, where ∂U is the boundary of U in X
>
> (b) there exists an effective bound $M \geq 1$, $\beta \in [0, \frac{1}{2})$, and a continuous function $\xi : [0,1] \mapsto \mathbb{R}$, and any $r \in \mathbb{N}$ such that $\rho(H^r(x, \lambda), H^r(y, \mu)) \leq M\beta^r[\rho(x, H(x, \lambda)) + \rho(y, H(y, \mu))] + |\xi(\lambda) - \xi(\mu)|$ for all $x, y \in \bar{U}$ and $\lambda, \mu \in [0,1]$
>
> If $H(\cdot, 0)$ has a r-fixed point in U, then $H(\cdot, \lambda)$ has a r-fixed point in U for every $\lambda \in [0,1]$

2.5 References

(1) Ampadu, Clement (2016). lulu.com, Characterization Theorems Inspired by the Hardy-Rogers Map I: Some Results in Metric Spaces. ISBN: 1365101185, 9781365101182

(2) S. Matthews, Partial metric topology, in: Proc. 8th Summer Conference on General Topology and Applications. Ann. New York Acad. Sci. 728 (1994), 183-197

(3) S. Oltra, O. Valero, Banach's fixed point theorem for partial metric spaces, Rend. Ist. Mat. Univ. Trieste. 36 (2004), 17-26

(4) Cristina Di Bari and Pasquale Vetro, Common Fixed Points for ψ-contractions on Partial Metric Spaces, Hacettepe Journal of Mathematics and Statistics Volume 42 (6) (2013), 591-598

Chapter 3

r-Best-Proximity Point Theorems for Weak Generalization of Higher-Order Hardy-Rogers Type Mapping in Partial Metric Spaces

3.1 Brief Summary

Abstract C.1 1

In Chapter 3 [Ampadu, Clement (2016). lulu.com, Characterization Theorems Inspired by the Hardy-Rogers Map I: Some Results in Metric Spaces. ISBN: 1365101185, 9781365101182] we obtained some higher-order best proximity point theorems for weak generalization of the higher-order Hardy-Rogers mapping with partial orders in metric spaces. In this chapter we obtain some higher-order best proximity point theorems for weak generalization of the higher-order Hardy-Rogers mapping in partial metric spaces. Following Proposition 3.1 [Dugundji, J, Granas, A: Weakly contractive mappings and elementary domain invariance theorem. Bull. Greek Math. Soc. 19, 141-151 (1978)] we use an equivalent form of Definition C.1 [Ampadu, Clement (2016). lulu.com, Characterization Theorems Inspired by the Hardy-Rogers Map I: Some Results in Metric Spaces. ISBN: 1365101185, 9781365101182] in the setting of this chapter.

3.2 Preliminaries

Definition C.1 1

A partial metric on a nonempty set X is a function $\rho : X \times X \mapsto \mathbb{R}^+$ such that for all $x, y, z \in X$

(a) $x = y$ iff $\rho(x, x) = \rho(x, y) = \rho(y, y)$

(b) $\rho(x, x) \leq \rho(x, y)$

(c) $\rho(x, y) = \rho(y, x)$

(d) $\rho(x, y) \leq \rho(x, z) + \rho(z, y) - \rho(z, z)$

Furthermore we say (X, ρ) is a partial metric space

Remark C.2 1

From (a) and (b) of the above definition, we note that $\rho(x,y) = 0$ implies $x = y$. However the converse is not necessarily true

Example C.3 1

Define $\rho : X \times X \mapsto \mathbb{R}^+$ by $\rho(x,y) = \max\{x,y\}$ for all $x, y \in \mathbb{R}^+$, then $(\mathbb{R}^+, \rho)$ is a partial metric space

Remark C.4 1

Each partial metric ρ on X generates a T_0 topology $T(\rho)$ on X which has as a base the family of open ρ-balls $\{B_\rho(x;\epsilon) : x \in X; \epsilon > 0\}$, where, $B_p(x,\epsilon) = \{y \in X : \rho(x,y) < \rho(x,x) + \epsilon\}$ for all $x \in X$ and $\epsilon > 0$

Definition C.5 1

A sequence $\{x_n\}$ in a partial metric space (X, ρ) converges to a point $x \in X$ iff $\rho(x,x) = \lim_{n \to \infty} \rho(x, x_n)$

Definition C.6 1

A sequence $\{x_n\}$ in a partial metric space (X, ρ) is called a Cauchy sequence if $\lim_{n,m \to \infty} \rho(x_n, x_m) < \infty$

Definition C.7 1

A partial metric space (X, ρ) is said to be complete if every Cauchy sequence $\{x_n\}$ in X converges, with respect to $T(\rho)$, to a point $x \in X$ such that $\rho(x,x) = \lim_{n,m \to \infty} \rho(x_n, x_m)$

Definition C.8 1

A mapping $f : X \mapsto X$ will be called r-continuous at $x_0 \in X$ if for every $\epsilon > 0$, there exists $\delta > 0$ such that $f^r(B(x_0, \delta)) \subseteq B(f^r(x_0), \epsilon)$ for any $r \in \mathbb{N}$

Remark C.9 1

If (X, ρ) is a partial metric space, then the function $d_p : X \times X \mapsto \mathbb{R}^+$ defined by $d_\rho(x,y) = 2\rho(x,y) - \rho(x,x) - \rho(y,y)$ is a metric on X

Remark C.10 1

[Matthews, SG: Partial metric topology. Ann. N.Y. Acad. Sci. 728, 183-197 (1994)] If a sequence converges in a partial metric space (X, ρ) with respect to $T(d_\rho)$, then it converges with respect to $T(\rho)$, but not conversely

Remark C.11 1

[Matthews, SG: Partial metric topology. Ann. N.Y. Acad. Sci. 728, 183-197 (1994)] A sequence $\{x_n\}$ in a partial metric space (X, ρ) is a Cauchy sequence iff it is a Cauchy sequence in the metric space (X, d_ρ)

Remark C.12 1

[Matthews, SG: Partial metric topology. Ann. N.Y. Acad. Sci. 728, 183-197 (1994)] A partial metric space (X, ρ) is complete iff the metric space (X, d_ρ) is complete

Remark C.13 1

[Matthews, SG: Partial metric topology. Ann. N.Y. Acad. Sci. 728, 183-197 (1994)] Given a sequence $\{x_n\}$ in a partial metric space (X, ρ) and $x \in X$, we have that $\lim_{n \to \infty} d_\rho(x, x_n) = 0$ iff $\rho(x,x) = \lim_{n \to \infty} \rho(x, x_n) = \lim_{n,m \to \infty} \rho(x_n, x_m)$

Definition C.14 1

Let (X, ρ) be a partial metric space. A map $T : X \mapsto X$ will be called a Hardy-Rogers type map if it satisfies $\rho(Tx, Ty) \leq k[\rho(x, Tx) + \rho(y, Ty) + \rho(x, Ty) + \rho(y, Tx) + \rho(x, y)]$ for all $x, y \in X$, where $k < \frac{1}{5}$ is nonnegative

Remark C.15 1

If X is a metric space in the above definition, that is, $\rho := d$, then we obtain Definition A.1 [Ampadu, Clement (2016). lulu.com, Characterization Theorems Inspired by the Hardy-Rogers Map I: Some Results in Metric Spaces. ISBN: 1365101185, 9781365101182]

The above definition can be stated in the following way

Definition C.16 1

Let (X, ρ) be a partial metric space. A map $T : X \mapsto X$ will be called a Hardy-Rogers type map if it satisfies $\rho(Tx, Ty) \leq \frac{\alpha}{5}[\rho(x, Tx) + \rho(y, Ty) + \rho(x, Ty) + \rho(y, Tx) + \rho(x, y)]$ for all $x, y \in X$, where $\alpha < 1$ is nonnegative

Remark C.17 1

If X is a metric space in the above definition, that is, $\rho := d$, then we obtain Definition A.1 [Ampadu, Clement (2016). lulu.com, Characterization Theorems Inspired by the Hardy-Rogers Map I: Some Results in Metric Spaces. ISBN: 1365101185, 9781365101182] in a different form

By Definition C.16, and taking inspiration from Proposition 3.1 [Dugundji, J, Granas, A: Weakly contractive mappings and elementary domain invariance theorem. Bull. Greek Math. Soc. 19, 141-151 (1978)], we have the following

Definition C.18 1

Let (X, ρ) be a partial metric space, and $T : X \mapsto X$. We will say T is a weakly Hardy-Rogers type map if there exists $\bar{\alpha} : X \times X \mapsto [0, 1)$ such that for every $0 \leq a \leq b$, $\theta(a, b) = \sup\{\bar{\alpha}(x, y) : a \leq \rho(x, y) \leq b\} < 1$; for all $x, y \in X$, $\rho(Tx, Ty) \leq \frac{\bar{\alpha}(x,y)}{5}[\rho(x, Tx) + \rho(y, Ty) + \rho(y, Tx) + \rho(x, Ty) + \rho(x, y)]$

From now on we concentrate on obtaining the higher-order version of the above definition when T is a non-self map. Let A and B be nonempty subsets of a partial metric space (X, ρ), and $T : A \mapsto B$. For any $r \in \mathbb{N}$, the r-best-proximity point problem is to find an element $x_0 \in A$ such that $\rho(x_0, T^r x_0) = \rho(A, B)$, where $\rho(A, B) = \inf\{\rho(x, y) : x \in A \text{ and } y \in B\}$. Since $\rho(x, T^r x) \geq \rho(A, B)$ for any $x \in A$, it follows that the optimal solution to this problem is one for which the value $\rho(A, B)$ is attained

Remark C.19 1

The 1-best proximity point problem has been explored. For examples, see [Gabeleh, M: Global optimal solutions of non-self mappings. U.P.B. Sci. Bull., Ser. A 75, 67-74 (2013); Zhang, J, Su, Y, Cheng, Q: A note on 'A best proximity point theorem for Geraghty-contractions'. Fixed Point Theory Appl. 2013, 99 (2013); Al-Thagafi, MA, Shahzad, N: Convergence and existence results for best proximity points. Nonlinear Anal. 70,3665-3671 (2009); Eldred, AA, Veeramani, P: Existence and convergence of best proximity points. J. Math. Anal. Appl. 323, 1001-1006 (2006); Alghamdi, MA, Shahzad, N, Vetro, F: Best proximity points for some classes of proximal contractions. Abstr. Appl. Anal. 2013, Article ID 713252 (2013)]

Notation C.20 1

Let A and B be nonempty subsets of a partial metric space (X, ρ). We denote by A_0 and B_0 the following sets: $A_0 = \{x \in A : \rho(x,y) = \rho(A,B) \text{ for some } y \in B\}$ and $B_0 = \{y \in B : \rho(x,y) = \rho(A,B) \text{ for some } x \in A\}$

Definition C.21 1

Let (A, B) be a pair of nonempty subsets of a partial metric space (X, ρ). A mapping $T: A \mapsto B$ will be called weakly Hardy-Rogers type provided that $\rho(Tx, Ty) \leq \frac{\bar{\alpha}(x,y)}{5}[\rho(x, Tx) + \rho(y, Ty) + \rho(x, Ty) + \rho(y, Tx) + \rho(x, y) - 4\rho(A, B)]$ for all $x, y \in A$, where the function $\bar{\alpha}: A \times A \mapsto [0, 1)$ satisfies for every $0 < a < b$, $\theta(a,b) = \sup\{\bar{\alpha}(x,y) : a \leq \rho(x,y) \leq b\} < 1$

Definition C.22 1

Let (A, B) be a pair of nonempty subsets of a partial metric space (X, ρ). A mapping $T: A \mapsto B$ will be called higher-order weakly Hardy-Rogers type provided that

$$\rho(T^r x, T^r y) \leq \sum_{q=0}^{r-1} \frac{\bar{\alpha}_q(x,y)}{5}[\rho(T^q x, T^{q+1} x) + \rho(T^q y, T^{q+1} y) + \rho(T^q x, T^{q+1} y)]$$
$$+ \sum_{q=0}^{r-1} \frac{\bar{\alpha}_q(x,y)}{5}[\rho(T^q y, T^{q+1} x) + \rho(T^q x, T^q y) - 4\rho(A, B)]$$

for all $x, y \in A$, where the function $\bar{\alpha}_q : A \times A \mapsto [0, 1)$ satisfies for every $0 < a < b$, $0 \leq q \leq r-1$, $\theta(a,b) = \sup\{\bar{\alpha}_q(x,y) : a \leq \rho(x,y) \leq b\} < 1$

Proposition C.23 1

Let (A, B) be a pair of nonempty subsets of a partial metric space (X, ρ), and let $T: A \mapsto B$ be a higher-order weakly Hardy-Rogers type map. For every pair $x \neq y \in X$, define

$$Q := Q(x,y) = \max_{0 \leq v \leq r-1} \zeta^{-v} \frac{\rho(T^v x, T^v y)}{\rho(x, Tx) + \rho(y, Ty) + \rho(x, Ty) + \rho(y, Tx) + \rho(x, y) - 4\rho(A, B)}$$

then

$$Q = \max_{n \in 0 \cup \mathbb{N}} \zeta^{-n} \frac{\rho(T^n x, T^n y)}{\rho(x, Tx) + \rho(y, Ty) + \rho(x, Ty) + \rho(y, Tx) + \rho(x, y) - 4\rho(A, B)}$$

where $\frac{\bar{\alpha}(x,y)}{5} := \zeta \in [0, \frac{1}{5})$, and $\bar{\alpha}: A \times A \mapsto [0, 1)$ satisfies for every $0 < a < b$, $\theta(a,b) = \sup\{\bar{\alpha}(x,y) : a \leq \rho(x,y) \leq b\} < 1$

> **Definition C.24 1**
>
> Let (A, B) be a pair of nonempty subsets of a partial metric space (X, ρ). A mapping $T : A \mapsto B$ will be called higher-order weakly Hardy-Rogers type if for all $x, y \in X$ and any $r \in \mathbb{N}$ it holds that $d(T^r x, T^r y) \leq Q\zeta^r [\rho(x, Tx) + \rho(y, Ty) + \rho(x, Ty) + \rho(y, Tx) + \rho(x, y) - 4\rho(A, B)]$, where Q and ζ are given by the previous proposition

> **Definition C.25 1**
>
> Let (A, B) be a pair of nonempty subsets of a partial metric space (X, ρ) with $A_0 \neq \emptyset$. Then the pair (A, B) is said to have the P-property iff, for any $x_1, x_2 \in A_0$ and $y_1, y_2 \in B_0$, $\rho(x_1, y_1) = \rho(A, B)$ and $\rho(x_2, y_2) = \rho(A, B)$ implies $\rho(x_1, x_2) = \rho(y_1, y_2)$

> **Lemma C.26 1**
>
> [Jingling Zhang and Yongfu Su, Best proximity point theorems for weakly contractive mapping and weakly Kannan mapping in partial metric spaces, Fixed Point Theory and Applications 2014, 2014:50] Let (X, ρ) be a partial metric space, then ρ is a continuous function, that is, for any $x_n, y_n, x, y \subseteq X$, if $x_n \to x$ and $y_n \to y$, then, $\rho(x_n, y_n) \to \rho(x, y)$ as $n \to \infty$

3.3 Main Results

Let Q^* be the bound obtained from Proposition C.23 when $\rho(A, B) = 0$, that is, $A = B := X$. It follows that T is a higher-order weakly Hardy-Rogers type self-map and taking inspiration from the authors in [Maryam A Alghamdi, Naseer Shahzad, and Oscar Valero, On fixed point theory in partial metric spaces, Fixed Point Theory and Applications 2012, 2012:175] we have the following

> **Theorem C.1 1**
>
> Let (X, ρ) be a complete partial metric space, and let $T : X \mapsto X$ be a mapping such that $\rho(T^r x, T^r y) \leq Q^* \zeta^r [\rho(x, Tx) + \rho(y, Ty) + \rho(x, Ty) + \rho(y, Tx) + \rho(x, y)]$ for all $x, y \in X$, where ζ is defined as in Proposition C.23. Then T has a unique r-fixed point $x^* \in X$ and the Picard sequence of iterates $\{T^{rn}(x_0)\}_{n \in \mathbb{N}}$ converges with respect to $T(d_\rho)$, for every $x_0 \in X$, to x^*. Moreover, $\rho(x^*, x^*) = 0$

> **Proof of Theorem C.1 1**
>
> Consider $x_0 \in X$ and define the Picard sequence of iterates by $x_n = T^r x_{n-1} = T^{rn} x_0$ for all $n \in \mathbb{N}$ and any $r \in \mathbb{N}$. Assume $x_n \neq x_{n+1}$ for all $n \in \mathbb{N}$, otherwise the existence of a r-fixed point is guaranteed. Now observe that
>
> $$\begin{aligned} \rho(x_n, x_{n+1}) &= \rho(T^r x_{n-1}, T^r x_n) \\ &\leq Q^* \zeta^r [\rho(x_{n-1}, T x_{n-1}) + \rho(x_n, T x_n) + \rho(x_{n-1}, T x_n) + \rho(x_n, T x_{n-1}) \\ &\quad + \rho(x_{n-1}, x_n)] \\ &= Q^* \zeta^r [\rho(x_{n-1}, x_n) + \rho(x_n, x_{n+1}) + \rho(x_{n-1}, x_{n+1}) + \rho(x_n, x_n) + \rho(x_{n-1}, x_n)] \\ &\leq Q^* \zeta^r [\rho(x_{n-1}, x_n) + \rho(x_n, x_{n+1}) + \rho(x_{n-1}, x_n) + \rho(x_n, x_{n+1}) - \rho(x_n, x_n) \\ &\quad + \rho(x_n, x_n) \\ &\quad + \rho(x_{n-1}, x_n)] \\ &= Q^* \zeta^r [3\rho(x_{n-1}, x_n) + 2\rho(x_n, x_{n+1})] \end{aligned}$$
>
> From the above one deduces that $\rho(x_n, x_{n+1}) \leq \Gamma \rho(x_{n-1}, x_n)$, where $\Gamma := \frac{3Q^* \zeta^r}{1 - 2Q^* \zeta^r} < 1$. Since $\Gamma < 1$, it follows that $\{\rho(x_n, x_{n+1})\}_{n \in \mathbb{N}}$ is a decreasing sequence and hence converges to a real number $p = \inf\{\rho(x_{n-1}, x_n) : n \in \mathbb{N}\}$. For the purpose of contradiction, assume that $p > 0$, then for all $n \in \mathbb{N}$, we have, $0 < p \leq \rho(x_n, x_{n+1}) \leq \rho(x_{n-1}, x_n) \leq \cdots \leq \rho(x_0, x_1)$ for all $n \in \mathbb{N}$. Since, $\theta = \theta(p, \rho(x_0, x_1))$, we obtain $\Gamma \leq \theta$, and thus, $p \leq \rho(x_n, x_{n+1}) \leq \Gamma \rho(x_n, x_{n-1}) \leq \theta^n \rho(x_0, x_1)$ for all $n \in \mathbb{N}$. Since $0 \leq \theta < 1$, we have, $\lim_{n \to \infty} \theta^n = 0$, and thus, $p = 0$, that is, $\lim_{n \to \infty} \rho(x_n, x_{n+1}) = 0$. Since $d_\rho(x_n, x_{n+1}) \leq 2\rho(x_n, x_{n+1})$ for all $n \in \mathbb{N}$, it follows that $\lim_{n \to \infty} d_\rho(x_n, x_{n+1}) = 0$. Furthermore, since $\rho(x_n, x_n) \leq \rho(x_n, x_{n+1})$ for all $n \in \mathbb{N}$, we have, $\lim_{n \to \infty} \rho(x_n, x_n) = 0$. For $k \in \mathbb{N}$, we have the following
>
> $$\begin{aligned} d_\rho(x_n, x_{n+k}) &\leq d_\rho(x_n, x_{n+1}) + \cdots + d_\rho(x_{n+k-1}, x_{n+k}) \\ &\leq 2\theta^n \rho(x_0, x_1) + \cdots + 2\theta^{n+k-1} \rho(x_0, x_1) \\ &= 2\theta^n \rho(x_0, x_1)[1 + \cdots + \theta^{k-1}] \\ &\leq \frac{2\theta^n}{1 - \theta} \rho(x_0, x_1) \end{aligned}$$
>
> where $\theta = \theta(0, \rho(x_0, x_1))$. This shows that $\{x_n\}$ is a Cauchy sequence in the metric space (X, d_ρ). Since (X, ρ) is complete, it follows that (X, d_ρ) is complete. So $\{x_n\}$ converges in the metric space (X, d_ρ), that is, $\lim_{n \to \infty} d_\rho(x_n, x^*) = 0$ for some $x^* \in X$. Moreover, $\rho(x^*, x^*) = \lim_{n \to \infty} \rho(x_n, x^*) = \lim_{n, m \to \infty} \rho(x_n, x_m)$. It follows that $\{x_n\}$ is a convergent sequence in (X, ρ) with respect to $T(d_\rho)$. Since $\{x_n\}$ is a Cauchy sequence in the metric space (X, d_ρ), we have, $\lim_{n, m \to \infty} d_\rho(x_n, x_m) = 0$. Moreover, since $\rho(x_n, x_n) \leq \rho(x_n, x_{n+1})$, we have, $\lim_{n \to \infty} \rho(x_n, x_n) = 0$. From the definition of d_ρ, we have, $\lim_{n, m \to \infty} \rho(x_n, x_m) = 0$, and thus, $\rho(x^*, x^*) = \lim_{n \to \infty} \rho(x_n, x^*) = \lim_{n, m \to \infty} \rho(x_n, x_m) = 0$. Now we show that $\rho(x^*, T^r x^*) = 0$. Suppose not, that is, $\rho(x^*, T^r x^*) > 0$, then we observe that,
>
> $$\begin{aligned} \rho(x^*, T^r x^*) &\leq \rho(x^*, T^r x_n) + \rho(T^r x_n, T^r x^*) - \rho(T^r x_n, T^r x_n) \\ &\leq \rho(x^*, x_{n+1}) + \rho(T^r x_n, T^r x^*) \\ &\leq \rho(x^*, x_{n+1}) + Q^* \zeta^r [\rho(x_n, T x_n) + \rho(x^*, T x^*) + \rho(x_n, T x^*) + \rho(x^*, T x_n) \\ &\quad + \rho(x_n, x^*)] \\ &= \rho(x^*, x_{n+1}) + Q^* \zeta^r [\rho(x_n, x_{n+1}) + \rho(x^*, T x^*) + \rho(x_n, T x^*) + \rho(x^*, x_{n+1}) \\ &\quad + \rho(x_n, x^*)] \\ &\leq \rho(x^*, x_{n+1}) + Q^* \zeta^r [\rho(x_n, x_{n+1}) + \rho(x^*, T x^*) + \rho(x_n, x^*) + \rho(x^*, T x^*) \\ &\quad - \rho(x^*, x^*) + \rho(x^*, x_{n+1}) + \rho(x_n, x^*)] \\ &\leq \rho(x^*, x_{n+1}) + Q^* \zeta^r [\rho(x_n, x_{n+1}) + \rho(x^*, T^r x^*) + \rho(x_n, x^*) + \rho(x^*, T^r x^*) \\ &\quad + \rho(x^*, x_{n+1}) + \rho(x_n, x^*)] \end{aligned}$$

Proof of Theorem C.1 continued 1

From the above one deduces that

$$(1-2Q^*\zeta^r)\rho(x^*, T^r x^*) \leq \rho(x^*, x_{n+1}) + Q^*\zeta^r[\rho(x_n, x_{n+1}) + \rho(x_n, x^*) + \rho(x^*, x_{n+1}) + \rho(x_n, x^*)]$$

Since $1 - 2Q^*\zeta^r \neq 0$, if we take limits in the above inequality, we get $\rho(x^*, T^r x^*) = 0$, which means $x^* = T^r x^*$, and existence of r-fixed point is proved. Finally, we show uniqueness of the r-fixed point. Suppose that z is another r-fixed point of T with $z \neq x^*$, it follows that $\rho(x^*, z) > 0$. Now observe that

$$\begin{aligned}\rho(x^*, z) &= \rho(T^r z, T^r x^*) \\ &\leq Q^*\zeta^r[\rho(x^*, Tx^*) + \rho(z, Tz) + \rho(x^*, Tz) + \rho(z, Tx^*) + \rho(z, x^*)] \\ &= Q^*\zeta^r[\rho(x^*, x^*) + \rho(z, z) + \rho(x^*, z) + \rho(z, x^*) + \rho(z, x^*)]\end{aligned}$$

Since $\rho(x^*, x^*), \rho(z, z) \leq \rho(x^*, z)$, it follows from the above inequality that, $\rho(x^*, z) \leq 5Q^*\zeta^r \rho(x^*, z)$. Moreover, $1 - 5Q^*\zeta^r \neq 0$, thus, $\rho(x^*, z) = 0$ which means $x^* = z$, a contradiction. Thus, uniqueness of the r-fixed point follows.

Now the main theorem is as follows

Theorem C.2 1

Let (A, B) be a pair of nonempty closed subsets of a complete partial metric space (X, ρ) such that $A_0 \neq \emptyset$. Let $T : A \mapsto B$ be a continuous higher-order weakly Hardy-Rogers type mapping. Suppose that $T^r(A_0) \subseteq B_0$ for any $r \in \mathbb{N}$ and the pair (A, B) has the P-property. Then T has a unique r-best proximity point $x^* \in A_0$ and the iterative sequence $\{x_{2k}\}_{k=0}^{\infty}$ defined by $x_{2k+1} = T^r x_{2k}$, $\rho(x_{2k+1}, x_{2k}) = \rho(A, B)$, $k = 0, 1, 2, \cdots$, and $r \in \mathbb{N}$, converges, with respect to $T(d_\rho)$, for every $x_0 \in A_0$, to x^*

Proof of Theorem C.2 1

At first we prove that B_0 is closed and $T^r(\bar{A}_0) \subseteq B_0$ for any $r \in \mathbb{N}$. Let $\{y_n\} \subseteq B_0$ be a sequence such that $y_n \to q \in B$. From the P-property we have that

$$d_\rho(y_n, y_m) = 2\rho(y_n, y_m) - \rho(y_n, y_n) - \rho(y_m, y_m)$$
$$= 2\rho(x_n, x_m) - \rho(x_n, x_n) - \rho(x_m, x_m)$$
$$= d_\rho(x_n, x_m)$$

So $d_\rho(y_n, y_m) \to 0$ implies $d_\rho(x_n, x_m) \to 0$ as $n, m \to \infty$, where $x_n, x_m \in A_0$ and $\rho(x_n, y_n) = \rho(A, B)$, $\rho(x_m, y_m) = \rho(A, B)$. It follows that $\{x_n\}$ is a Cauchy sequence in (X, d_ρ), and thus converges to a point $p \in A$. By continuity of partial metric ρ, we have $\rho(p, q) = \rho(A, B)$, that is, $q \in B_0$, and hence B_0 is closed with respect to (X, d_ρ). Now let $\bar{A}_0$ be the closure of A_0 in the metric space (X, d_ρ). If $x \in \bar{A}_0 \setminus A_0$, then there exists a sequence $\{x_n\} \subseteq A_0$ such that $x_n \to x$. By the r-continuity of T and the closedness of B_0, we have, $T^r x = \lim_{n \to \infty} T^r x_n \in B_0$, that is, $T^r(\bar{A}_0) \subseteq B_0$. Now define $P_{A_0} T : \bar{A}_0 \mapsto \bar{A}_0$ by $P_{A_0} y = \{x \in A_0 : \rho(x, y) = \rho(A, B)\}$. Since the pair (A, B) has the P-property, we have,

$$\rho((P_{A_0}T)^r x_1, (P_{A_0}T)^r x_2) = \rho(T^r x_1, T^r x_2)$$
$$\leq Q\zeta^r[\rho(x_1, Tx_1) + \rho(x_2, Tx_2) + \rho(x_1, Tx_2) + \rho(x_2, Tx_1)$$
$$+ \rho(x_1, x_2) - 4\rho(A, B)]$$
$$\leq Q\zeta^r[\rho(x_1, P_{A_0}Tx_1) + \rho(P_{A_0}Tx_1, Tx_1) + \rho(x_2, P_{A_0}Tx_2)$$
$$+ \rho(P_{A_0}Tx_2, Tx_2)$$
$$+ \rho(x_1, P_{A_0}Tx_2) + \rho(P_{A_0}Tx_2, Tx_2) + \rho(x_2, P_{A_0}Tx_1)$$
$$+ \rho(P_{A_0}Tx_1, Tx_1)$$
$$+ \rho(x_1, x_2) - 4\rho(A, B)]$$
$$= Q\zeta^r[\rho(x_1, P_{A_0}Tx_1) + \rho(x_2, P_{A_0}Tx_2) + \rho(x_1, P_{A_0}Tx_2)$$
$$+ \rho(x_2, P_{A_0}Tx_1)$$
$$+ \rho(x_1, x_2)]$$

for any $x_1, x_2 \in \bar{A}_0$. It follows that $P_{A_0}T : \bar{A}_0 \mapsto \bar{A}_0$ is a higher-order weakly Hardy-Rogers type map from a complete partial metric subspace $\bar{A}_0$ into itself. By the previous theorem, $P_{A_0}T : \bar{A}_0 \mapsto \bar{A}_0$ has a unique r-fixed point x^*, that is, $(P_{A_0}T)^r x^* = x^* \in A_0$, which implies that $\rho(x^*, T^r x^*) = \rho(A, B)$. Moreover, $\{(P_{A_0}T)^{rn} x_0\}_{n \in \mathbb{N}}$ converges, with respect to $T(d_\rho)$, for every $x_0 \in A_0$, to x^*. Since the iteration sequence $\{x_{2k}\}_{k=0}^\infty$ defined by $x_{2k+1} = T^r x_{2k}$, $\rho(x_{2k+1}, x_{2k}) = \rho(A, B)$, $k = 0, 1, 2, \cdots$, and $r \in \mathbb{N}$ is exactly the subsequence of $\{x_n\}$. It follows that it converges, for every $x_0 \in A_0$, to x^*.

3.4 Exercises

Exercise C.1 1

Using the techniques developed in this chapter, obtain the higher-order version of Corollary 17[Alghamdi, MA, Shahzad, N, Valero, O: On fixed point theory in partial metric spaces. Fixed Point Theory Appl. 2012, 175 (2012)]

Exercise C.2 1

Using the techniques developed in this chapter, obtain the higher-order version of Corollary 18 [Alghamdi, MA, Shahzad, N, Valero, O: On fixed point theory in partial metric spaces. Fixed Point Theory Appl. 2012, 175 (2012)]

> **Exercise C.3 1**
>
> Using Exercise C.1, obtain the higher-order version of Theorem 2.5 [Jingling Zhang and Yongfu Su, Best proximity point theorems for weakly contractive mapping and weakly Kannan mapping in partial metric spaces, Fixed Point Theory and Applications 2014, 2014:50]

> **Exercise C.4 1**
>
> Using Exercise C.2, obtain the higher-order version of Theorem 2.6 [Jingling Zhang and Yongfu Su, Best proximity point theorems for weakly contractive mapping and weakly Kannan mapping in partial metric spaces, Fixed Point Theory and Applications 2014, 2014:50]

3.5 References

(1) Ampadu, Clement (2016). lulu.com, Characterization Theorems Inspired by the Hardy-Rogers Map I: Some Results in Metric Spaces. ISBN: 1365101185, 9781365101182

(2) Matthews, SG: Partial metric topology. Ann. N.Y. Acad. Sci. 728, 183-197 (1994)

(3) Dugundji, J, Granas, A: Weakly contractive mappings and elementary domain invariance theorem. Bull. Greek Math. Soc. 19, 141-151 (1978)

(4) Gabeleh, M: Global optimal solutions of non-self mappings. U.P.B. Sci. Bull., Ser. A 75, 67-74 (2013)

(5) Zhang, J, Su, Y, Cheng, Q: A note on 'A best proximity point theorem for Geraghty-contractions'. Fixed Point Theory Appl. 2013, 99 (2013)

(6) Al-Thagafi, MA, Shahzad, N: Convergence and existence results for best proximity points. Nonlinear Anal. 70,3665-3671 (2009)

(7) Eldred, AA, Veeramani, P: Existence and convergence of best proximity points. J. Math. Anal. Appl. 323, 1001-1006 (2006)

(8) Alghamdi, MA, Shahzad, N, Vetro, F: Best proximity points for some classes of proximal contractions. Abstr. Appl. Anal. 2013, Article ID 713252 (2013)

(9) Jingling Zhang and Yongfu Su, Best proximity point theorems for weakly contractive mapping and weakly Kannan mapping in partial metric spaces, Fixed Point Theory and Applications 2014, 2014:50

(10) Maryam A Alghamdi, Naseer Shahzad, and Oscar Valero, On fixed point theory in partial metric spaces, Fixed Point Theory and Applications 2012, 2012:175

Chapter 4

Graphic Weak Ψ-Higher Order Contractions with Application to Integral Equations

4.1 Brief Summary

Abstract D.1 1

Using the notion of triangular r-α-admissibility, Chapter 5 [Ampadu, Clement (2016). lulu.com, Characterization Theorems Inspired by the Hardy-Rogers Map I: Some Results in Metric Spaces. ISBN: 1365101185, 9781365101182] we introduce a concept of modified-modified weak α-ψ-higher-order contractive mappings and establish some fixed point results for such mappings defined on ordinary as well as ordered partial metric spaces. As application we introduce a concept of modified weak ψ-graphic higher-order contractive mappings and obtain some fixed point results. Application to integral equations are also considered

4.2 Preliminaries

Definition D.1 1

A partial metric on a nonempty set X is a function $\rho : X \times X \mapsto \mathbb{R}^+$ such that for all $x, y, z \in X$

(a) $x = y$ iff $\rho(x,x) = \rho(x,y) = \rho(y,y)$

(b) $\rho(x,x) \leq \rho(x,y)$

(c) $\rho(x,y) = \rho(y,x)$

(d) $\rho(x,y) \leq \rho(x,z) + \rho(z,y) - \rho(z,z)$

Furthermore we say (X, ρ) is a partial metric space

Remark D.2 1

From (a) and (b) of the above definition, we note that $\rho(x,y) = 0$ implies $x = y$. However the converse is not necessarily true

Example D.3 1

Define $\rho : X \times X \mapsto \mathbb{R}^+$ by $\rho(x,y) = \max\{x,y\}$ for all $x,y \in \mathbb{R}^+$, then $(\mathbb{R}^+, \rho)$ is a partial metric space

Remark D.4 1

Each partial metric ρ on X generates a T_0 topology $T(\rho)$ on X which has as a base the family of open ρ-balls $\{B_\rho(x;\epsilon) : x \in X; \epsilon > 0\}$, where, $B_p(x,\epsilon) = \{y \in X : \rho(x,y) < \rho(x,x) + \epsilon\}$ for all $x \in X$ and $\epsilon > 0$

Definition D.5 1

A sequence $\{x_n\}$ in a partial metric space (X,ρ) converges to a point $x \in X$ iff $\rho(x,x) = \lim_{n \to \infty} \rho(x, x_n)$

Definition D.6 1

A sequence $\{x_n\}$ in a partial metric space (X,ρ) is called a Cauchy sequence if $\lim_{n,m \to \infty} \rho(x_n, x_m) < \infty$

Definition D.7 1

A partial metric space (X,ρ) is said to be complete if every Cauchy sequence $\{x_n\}$ in X converges, with respect to $T(\rho)$, to a point $x \in X$ such that $\rho(x,x) = \lim_{n,m \to \infty} \rho(x_n, x_m)$

Definition D.8 1

A sequence $\{x_n\}$ in a partial metric space (X,ρ) is called 0-Cauchy if $\lim_{n,m \to \infty} \rho(x_n, x_m) = 0$

Definition D.9 1

We say that (X,ρ) is 0-Complete if every 0-Cauchy sequence in X converges, with respect to $T(\rho)$, to a point $x \in X$ such that $\rho(x,x) = 0$

Example D.10 1

Let $\mathbb{Q}$ denote the set of rational numbers, then, the partial metric space $(\mathbb{Q} \cap [0,\infty), \max\{x,y\})$ is 0-complete but not complete

Remark D.11 1

If (X,ρ) is a partial metric space, then the function $\rho^s : X \times X \mapsto \mathbb{R}^+$ defined by $\rho^s(x,y) = 2\rho(x,y) - \rho(x,x) - \rho(y,y)$ is a metric on X

Remark D.12 1

[Matthews, SG: Partial metric topology. Ann. N.Y. Acad. Sci. 728, 183-197 (1994)] A partial metric space (X,ρ) is complete iff the metric space (X, ρ^s) is complete

Remark D.13 1

[Matthews, SG: Partial metric topology. Ann. N.Y. Acad. Sci. 728, 183-197 (1994)] Given a sequence $\{x_n\}$ in a partial metric space (X,ρ) and $z \in X$, we have that $\lim_{n\to\infty} \rho^s(z,x_n) = 0$ iff $\rho(z,z) = \lim_{n\to\infty} \rho(z,x_n) = \lim_{n,m\to\infty} \rho(x_n,x_m)$

Definition D.14 1

(Definition E.4, [Ampadu, Clement (2016). lulu.com, Characterization Theorems Inspired by the Hardy-Rogers Map I: Some Results in Metric Spaces. ISBN: 1365101185, 9781365101182]) Let T be a self-mapping on X and $\alpha : X \times X \mapsto [0,\infty)$. One says that T is an r-α-admissible mapping if $x,y \in X$, $\alpha(x,y) \geq 1$ implies $\alpha(T^r x, T^r y) \geq 1$

Notation D.15 1

Ψ will denote the family of nondecreasing functions $\psi : [0,\infty) \mapsto [0,\infty)$ such that $\sum_{n=1}^{\infty} \psi^n(t) < \infty$ for all $t > 0$, where ψ^n denotes the nth iterate of ψ

Taking inspiration from [P. Salimi, A. Latif, and N. Hussain, "Modified $\alpha-\psi$-contractive mappings with applications," Fixed Point Theory and Applications, vol. 2013, article 151, 2013] we introduce the following

Definition D.16 1

Let T be a self-mapping on X and $\alpha, \eta : X \times X \mapsto [0,\infty)$. We will say that T is r-α-admissible with respect to η if $x,y \in X$, $\alpha(x,y) \geq \eta(x,y)$ implies $\alpha(T^r x, T^r y) \geq \eta(T^r x, T^r y)$

Remark D.17 1

If $\eta(x,y) = 1$ in the definition above, then we recover Definition D.14

Remark D.18 1

If we take $\alpha(x,y) = 1$ in Definition D.16, then we say that T is r-η-sub-admissible

Definition D.19 1

(Definition E.6, [Ampadu, Clement (2016). lulu.com, Characterization Theorems Inspired by the Hardy-Rogers Map I: Some Results in Metric Spaces. ISBN: 1365101185, 9781365101182]) Let T be a self-mapping on X and $\alpha : X \times X \mapsto (-\infty, \infty)$. One says that T is triangular-r-α-admissible mapping if in addition to being r-α-admissible it satisfies $\alpha(x,z) \geq 1$ and $\alpha(z,y) \geq 1$ imply $\alpha(x,y) \geq 1$

Lemma D.20 1

Let $T : X \mapsto X$ be a triangular r-α-admissible mapping. Assume there exists $x_0 \in X$ such that $\alpha(x_0, T^r x_0) \geq 1$ for any $r \in \mathbb{N}$. Define a sequence $\{x_n\}$ by $x_n = T^r x_{n-1} = T^{rn} x_0$. Then we have $\alpha(x_m, x_n) \geq 1$ for all $n, m \in \mathbb{N}$ with $m < n$

Proof of Lemma D.20 1

Lemma E.1 [Ampadu, Clement (2016). lulu.com, Characterization Theorems Inspired by the Hardy-Rogers Map I: Some Results in Metric Spaces. ISBN: 1365101185, 9781365101182])]

CHAPTER 4. GRAPHIC WEAK Ψ-HIGHER ORDER CONTRACTIONS WITH APPLICATION TO INTEGRAL EQUATIONS

Definition D.21 1

Let (X, ρ) be a partial metric space and $T : X \mapsto X$ and $\alpha : X \times X \mapsto [0, \infty)$ be two mappings. If there exists an upper semi-continuous from the right nondecreasing function $\psi : \mathbb{R}^+ \mapsto \mathbb{R}^+$ with $\psi(t) < t$ for all $t > 0$ such that $x, y \in X$, $\alpha(x, y) \geq 1$ implies $\rho(T^r x, T^r) \leq \psi(M_\rho(x, y))$, where

$$M_\rho(x, y) = \max\{\rho(x, y), \rho(x, Tx), \rho(x, Ty), \rho(y, Tx), \rho(y, Ty)\}$$

, then we will say T is a modified-modified weak $\alpha - \psi$-higher-order contractive mapping.

Definition D.22 1

Let (X, ρ) be a partial metric space. Let $\alpha : X \times X \mapsto (-\infty, \infty)$ and $T : X \mapsto X$. We will say that T is an r-α continuous function on (X, ρ^s) if for a given x and $\{x_n\}$ in X, $x_n \to x$ as $n \to \infty$, $\alpha(x_n, x_{n+1}) \geq 1$ for all $n \in \mathbb{N}$ imply $T^r x_n \to T^r x$ for any $r \in \mathbb{N}$

Example D.23 1

Let $X = [0, \infty)$ and $\rho(x, y) = \max\{x, y\}$. Assume that $T : X \mapsto X$ and $\alpha : X^2 \mapsto \mathbb{R}$ are defined by

$$T^r(x) = \begin{cases} 3^r x & x \in [0, 1] \\ 5^r x + \frac{5^r - 1}{4} & x \in (1, \infty) \end{cases}$$

and

$$\alpha(x, y) = \begin{cases} 1 & x \in [0, 1] \\ -1 & x \notin [0, 1] \end{cases}$$

Note that T is not r-continuous, but is r-α continuous on (X, ρ^s). Indeed, if $\lim_{n \to \infty} x_n = x$ and $\alpha(x_n, x_{n+1}) \geq 1$, then $x_n \in [0, 1]$ and so $\lim_{n \to \infty} T^r x_n = \lim_{n \to \infty} 3^r x_n = 3^r x = T^r x$

Remark D.24 1

For some basic notions and notations in graph theory that would be useful in the sequel, the reader should see [M.Abbas and T.Nazir, "Common fixed point of a power graphic contraction pair in partial metric spaces endowed with a graph," Fixed Point Theory and Applications, vol. 2013, article 20, 2013; J. Jachymski, "The contraction principle for mappings on a metric space with a graph," Proceedings of the American Mathematical Society, vol. 136, no. 4, pp. 1359–1373, 2008; I. Beg, A. R. Butt, and S. Radojevic, "The contraction principle for set valued mappings on a metric space with a graph," Computers and Mathematics with Applications, vol. 60, no. 5, pp. 1214–1219, 2010; F. Bojor, "Fixed point theorems for Reich type contractions on metric spaces with a graph," Nonlinear Analysis: Theory, Methods and Applications, vol. 75, no. 9, pp. 3895–3901, 2012]

Taking inspiration from [J. Jachymski, "The contraction principle for mappings on a metric space with a graph," Proceedings of the American Mathematical Society, vol. 136, no. 4, pp. 1359–1373, 2008] we introduce the following

Definition D.25 1

A map $T : X \mapsto X$ will be called r-G-continuous if given $x \in X$ and a sequence $\{x_n\}$, $\lim_{n \to \infty} x_n = x$, $(x_n, x_{n+1}) \in E(G)$ for all $n \in \mathbb{N}$ imply $T^r x_n \to T^r x$ for every $r \in \mathbb{N}$

Definition D.26 1

Let (X, ρ) be a partial metric space endowed with a graph G and $T : X \mapsto X$ be a self-map. If there exists an upper semi-continuous from the right function $\psi : \mathbb{R}^+ \mapsto \mathbb{R}^+$ with $\psi(t) < t$ for all $t > 0$ such that

(a) for all $x, y \in X$, $(x, y) \in E(G)$ imply $(T^r x, T^r y) \in E(G)$

(b) for all $x, y \in X$, $(x, y) \in E(G)$ imply $\rho(T^r x, T^r y) \leq \psi(M_\rho(x, y))$, where

$$M_\rho(x, y) = \max\{\rho(x, y), \rho(x, Tx), \rho(x, Ty), \rho(y, Tx), \rho(y, Ty)\}$$

Then we will say that T is a modified weak ψ-graphic higher-order contraction mapping

Definition D.27 1

Let $(X, \rho, \preceq)$ be a partially ordered partial metric space endowed with a graph G and $T : X \mapsto X$. If there exists an upper semi-continuous from the right function $\psi : \mathbb{R}^+ \mapsto \mathbb{R}^+$ with $\psi(t) < t$ for all $t > 0$ such that for all $x, y \in X$

(a) $x, y \in E(G)$ with $x \preceq y$ implies $(T^r x, T^r y) \in E(G)$, where $T^r x \preceq T^r y$

(b) $x, y \in E(G)$ with $x \preceq y$ implies $\rho(T^r x, T^r y) \leq \psi(M_\rho(x, y))$, where

$$M_\rho(x, y) = \max\{\rho(x, y), \rho(x, Tx), \rho(x, Ty), \rho(y, Tx), \rho(y, Ty)\}$$

Then we will say T is an ordered modified weak ψ-graphic higher-order contraction

4.3 Main Results

Theorem D.1 1

Let (X, ρ) be a 0-complete partial metric space and T be a modified-modified α-ψ-higher-order contractive mapping and a triangular r-α admissible mapping from (X, ρ) into itself. Suppose that the following assertions hold:

(a) there exists $x_0 \in X$ such that $\alpha(x_0, T^r x_0) \geq 1$

(b) T is an r-α-continuous function on (X, ρ^s)

Then, T has r-fixed point

Proof of Theorem D.1 1

Let $x_0 \in X$ be such that $\alpha(x_0, T^r x_0) \geq 1$. Define a sequence $\{x_n\}$ in X by $x_n = T^{rn} x_0 = T^r x_{n-1}$ for all $n \in \mathbb{N}$. Then by Lemma D.20, we have, $\alpha(x_m, x_n) \geq 1$ for all $m, n \in \mathbb{N}$ with $m < n$. If $x_{n+1} = x_n$ for some $n \in \mathbb{N}$, then $x = x_n$ is a r-fixed point of T. Thus we assume that $x_{n+1} \neq x_n$ for all $n \in \mathbb{N}$. Now observe that $\rho(T^r x_{n-1}, T^r x_n) \leq \psi(M_\rho(x_{n-1}, x_n))$. On the other hand,

$$\max\{\rho(x_{n-1}, x_n), \rho(x_n, x_{n+1})\} \leq M_\rho(x_{n-1}, x_n)$$
$$= \max\{\rho(x_{n-1}, x_n), \rho(x_{n-1}, Tx_{n-1}), \rho(x_n, Tx_n),$$
$$\rho(x_{n-1}, Tx_n), \rho(x_n, Tx_{n-1})\}$$
$$= \max\{\rho(x_{n-1}, x_n), \rho(x_n, x_{n+1}), \rho(x_{n-1}, x_{n+1}), \rho(x_n, x_n)\}$$
$$\leq \max\{\rho(x_{n-1}, x_n), \rho(x_n, x_{n+1}), \rho(x_{n-1}, x_{n+1}), \rho(x_{n-1}, x_n)\}$$
$$= \max\{\rho(x_{n-1}, x_n), \rho(x_n, x_{n+1})\}$$

If $\max\{\rho(x_{n-1}, x_n), \rho(x_n, x_{n+1})\} = \rho(x_n, x_{n+1})$, then we get a contradiction, thus,

$$\rho(x_n, x_{n+1}) = \rho(T^r x_{n-1}, T^r x_n) \leq \psi(\rho(x_{n-1}, x_n)) < \rho(x_{n-1}, x_n)$$

It follows that $\{\rho(x_{n-1}, x_n)\}$ is a decreasing sequence, and there exists $s \geq 0$ such that $\lim_{n\to\infty} \rho(x_n, x_{n+1}) = \lim_{n\to\infty} \psi(\rho(x_{n-1}, x_n)) = s$. Now we show that $s = 0$. If $s > 0$, then, $s = \lim_{n\to\infty} \sup \psi(\rho(x_{n-1}, x_n)) \leq \psi(s) < s$, which is a contradiction. Hence $\lim_{n\to\infty} \rho(x_n, x_{n+1}) = 0$. Now we prove that $\{x_n\}$ is a 0-Cauchy sequence. Suppose not, then there exists $\epsilon > 0$ and sequences $\{m(k)\}$ and $\{n(k)\}$ such that, for all positive integers k, $n(k) > m(k) > k$, $\rho(x_{n(k)}, x_{m(k)}) \geq \epsilon$, and $\rho(x_{n(k)}, x_{m(k)-1}) < \epsilon$. Now for all $k \in \mathbb{N}$, we have,

$$\epsilon \leq \rho(x_{n(k)}, x_{m(k)})$$
$$\leq \rho(x_{n(k)}, x_{m(k)-1}) + \rho(x_{m(k)-1}, x_{m(k)})$$
$$< \epsilon + \rho(x_{m(k)-1}, x_{m(k)})$$

Taking limits in the above inequality and using the fact that $\lim_{n\to\infty} \rho(x_n, x_{n+1}) = 0$, we deduce that $\lim_{k\to\infty} \rho(x_{n(k)}, x_{m(k)}) = \epsilon$. Now observe that,

$$\rho(x_{n(k)}, x_{m(k)}) \leq \rho(x_{m(k)}, x_{m(k)+1})$$
$$+ \rho(x_{m(k)+1}, x_{n(k)+1}) + \rho(x_{n(k)+1}, x_{n(k)})$$

and

$$\rho(x_{n(k)+1}, x_{m(k)+1}) \leq \rho(x_{m(k)}, x_{m(k)+1})$$
$$+ \rho(x_{m(k)}, x_{n(k)}) + \rho(x_{n(k)+1}, x_{n(k)})$$

Since $\lim_{n\to\infty} \rho(x_n, x_{n+1}) = 0$ and $\lim_{k\to\infty} \rho(x_{n(k)}, x_{m(k)}) = \epsilon$, we deduce from the two chain of inequalities immediately above that, $\lim_{k\to\infty} \rho(x_{n(k)+1}, x_{m(k)+1}) = \epsilon$. Now since

$$\epsilon \leq \rho(x_{n(k)}, x_{m(k)})$$
$$\leq \rho(x_{n(k)}, x_{m(k)+1}) + \rho(x_{m(k)+1}, x_{m(k)})$$

it follows upon using the fact that $\lim_{n\to\infty} \rho(x_n, x_{n+1}) = 0$ and $\lim_{k\to\infty} \rho(x_{n(k)}, x_{m(k)}) = \epsilon$, that, $\lim_{k\to\infty} \rho(x_{n(k)}, x_{m(k)+1}) = \epsilon$. Similarly, we have, $\lim_{k\to\infty} \rho(x_{m(k)}, x_{n(k)+1}) = \epsilon$.

Proof of Theorem D.1 Continued 1

Finally, with

$$M_\rho(x_{n(k)}, x_{m(k)}) = \max\{\rho(x_{n(k)}, x_{m(k)}), \rho(x_{n(k)}, x_{n(k)+1}), \rho(x_{m(k)}, x_{m(k)+1}),$$
$$\rho(x_{n(k)}, x_{m(k)+1}), \rho(x_{m(k)}, x_{m(k)+1})\}$$

and using the facts that $\lim_{n\to\infty} \rho(x_n, x_{n+1}) = 0$, $\lim_{k\to\infty} \rho(x_{n(k)}, x_{m(k)}) = \epsilon$, $\lim_{k\to\infty} \rho(x_{n(k)}, x_{m(k)+1}) = \epsilon$, $\lim_{k\to\infty} \rho(x_{n(k)+1}, x_{m(k)+1}) = \epsilon$, $\lim_{k\to\infty} \rho(x_{m(k)}, x_{n(k)+1}) = \epsilon$, we get upon taking the limit supremum in, $\rho(x_{n(k)+1}, x_{m(k)+1}) = \rho(T^r x_{n(k)}, T^r x_{m(k)}) \leq \psi(M_\rho(x_{n(k)}, x_{m(k)}))$, that $\epsilon \leq \psi(\epsilon) < \epsilon$, which is a contradiction. Hence $\{x_n\}$ is a 0-Cauchy sequence. Since T is r-α-continuous on (X, ρ^s), $x_n \to z$ as $n \to \infty$ and $\alpha(x_n, x_{n+1}) \geq 1$, then we have, $T^r z = \lim_{n\to\infty} T^r x_n = \lim_{n\to\infty} x_{n+1} = z$. So z is a r-fixed point of T

Theorem D.2 1

Let (X, ρ) be a 0-complete partial metric space and $T : X \mapsto X$ be a modified-modified weak α-ψ-higher-order contraction and an r-α admissible mapping. Suppose the following assertions hold

(a) there exists $x_0 \in X$ such that $\alpha(x_0, T^r x_0) \geq 1$

(b) $\alpha(x, x) \geq 1$ for all $x \in X$ and if $\{x_n\}$ is a sequence in X such that $\alpha(x_n, x_{n+1}) \geq 1$ for all $n \in \mathbb{N}$ and $x_n \to x$ as $n \to \infty$, then $\alpha(x_n, x) \geq 1$ for all $n \in \mathbb{N}$

Then, T has a r-fixed point

Proof of Theorem D.2 1

(b) of the theorem implies second part of Definition D.19. Indeed if $\alpha(x, y) \geq 1$ and $\alpha(y, z) \geq 1$, then applying (b) of theorem to $\{x_n\}$ defined by $x_1 := x$, $x_2 := y$, $x_n := z$ for $n \geq 3$, we get $\alpha(x_n, z) \geq 1$ for all $n \in \mathbb{N}$, and hence $\alpha(x, z) \geq 1$. Thus, as in the previous theorem, we obtain a 0-Cauchy sequence $\{x_n\}$ such that $x_n \to z$ as $n \to \infty$. Since $\alpha(x_n, x_{n+1}) \geq 1$ for all $n \in \mathbb{N}$ and $\lim_{n\to\infty} x_n = z$, then from (b) of the theorem we have $\alpha(x_n, z) \geq 1$ for all $n \in \mathbb{N}$. Now observe that with

$$M_\rho(x_n, z) = \max\{\rho(x_n, z), \rho(x_n, x_{n+1}), \rho(z, Tz), \rho(x_n, Tz), \rho(z, x_{n+1})\}$$
$$\leq \max\{\rho(x_n, z), \rho(x_n, x_{n+1}), \rho(z, T^r z), \rho(x_n, T^r z), \rho(z, x_{n+1})\}$$

and taking limit supremum as $n \to \infty$ in, $\rho(x_{n+1}, T^r z) \leq \psi(M_\rho(x_n, z))$, we deduce that, $\rho(z, T^r z) \leq \psi(\rho(z, T^r z)) < \rho(z, T^r z)$, which is a contradiction. Hence, $\rho(z, T^r z) = 0$, that is, $z = T^r z$

> **Example D.3 1**
>
> Let $X = [0, \infty)$ be endowed with the partial metric $\rho(x,y) = \max\{x,y\}$ for all $x, y \in X$, and let $T : X \mapsto X$ be defined for any $r \in \mathbb{N}$ by,
>
> $$T^r(x) = \begin{cases} \frac{x}{2^r} & x \in [0,1] \\ x & x \in (1, \infty) \end{cases}$$
>
> Define $\alpha : X \times X \mapsto (-\infty, \infty)$ and $\psi : [0, \infty) \mapsto [0, \infty)$ by
>
> $$\alpha(x,y) = \begin{cases} 2 & x, y \in [0,1] \ or \ x = y \\ -2 & otherwise \end{cases}$$
>
> and $\psi(t) = \frac{t}{2}$. Note that (X, ρ) is a 0-complete partial metric space. We show that T is a triangular r-α-admissible mapping. Let $x, y \in X$, if $\alpha(x,y) \geq 1$, then $x, y \in [0,1]$ or $x = y$. On the other hand for all $x \in [0,1]$, we have, $T^r(x) \leq 1$ for any $r \in \mathbb{N}$. It follows that $\alpha(T^r x, T^r y) \geq 1$. Also if $\alpha(x,z) \geq 1$ and $\alpha(z,y) \geq 1$, then $x,y,z \in [0,1]$, that is, $\alpha(x,y) \geq 1$. Hence the assertion holds. In reason of the above arguments, $\alpha(0, T^r 0) \geq 1$ for any $r \in \mathbb{N}$. Now if $\{x_n\}$ is a sequence in X such that $\alpha(x_n, x_{n+1}) \geq 1$ for all $n \in \mathbb{N} \cup \{0\}$ and $x_n \to x$ as $n \to \infty$, then $\{x_n\} \subset [0,1]$, and hence $x \in [0,1]$. It follows that $\alpha(x_n, x) \geq 1$ for all $n \in \mathbb{N}$. Let $\alpha(x,y) \geq 1$, then $x, y \in [0,1]$, and we have,
>
> $$\rho(T^r x, T^r y) = \frac{1}{2^r} \max\{x,y\}$$
> $$\leq \frac{1}{2} \max\{x,y\}$$
> $$\leq \frac{1}{2} M_\rho(x,y)$$
> $$= \psi(M_\rho(x,y))$$
>
> that is, $\alpha(x,y) \geq 1$ implies $\rho(T^r x, T^r y) \leq \psi(M_\rho(x,y))$. Hence all the conditions of the previous theorem are satisfied and zero is the r-fixed point of T

As a consequence of the above theorems, we have the following

> **Corollary D.4 1**
>
> Let (X, ρ) be a 0-complete partial metric space and $T : X \mapsto X$ be a given mapping satisfying $\rho(T^r x, T^r y) \leq Q^* \zeta^r [\rho(x, Tx) + \rho(y, Ty) + \rho(x, Ty) + \rho(y, Tx) + \rho(x,y)]$ for all $x, y \in X$, where Q^* is a certain modification on Q, and Q and ζ are given by Proposition 1.22[Ampadu, Clement (2016):Ψ-Higher-Order Contractions and Some Common r-Fixed Point Theorems in 0-Complete Partial Metric Spaces. Unpublished]. Then T has a unique r-fixed point.

> **Remark D.5 1**
>
> The above Corollary can also be realized by taking f to be the identity in Theorem 2.4 [Ampadu, Clement (2016):Ψ-Higher-Order Contractions and Some Common r-Fixed Point Theorems in 0-Complete Partial Metric Spaces. Unpublished]

> **Theorem D.6 1**
>
> Let $(X, \rho, \preceq)$ be a partially ordered 0-complete partial metric space and $T : X \mapsto X$ be an r-increasing self-mapping such that $x_0 \preceq T^r x_0$ for some $x_0 \in X$ and any $r \in \mathbb{N}$. Assume that $\rho(T^r x, T^r y) \leq \psi(M_\rho(x,y))$ holds for all $x, y \in X$ and any $r \in \mathbb{N}$ with $x \preceq y$. If T is a r-continuous mapping on (X, ρ^s), then T has a r-fixed point.

Proof of Theorem D.6 1

Define $\alpha : X \times X \mapsto (-\infty, \infty)$ by

$$\alpha(x,y) = \begin{cases} 1 & if\ x \preceq y \\ 0 & otherwise \end{cases}$$

At first we prove that T is a triangular r-α-admissible mapping. Let $\alpha(x,y) \geq 1$, then $x \preceq y$. As T is an r-increasing mapping we have $T^r x \preceq T^r y$, that is, $\alpha(T^r x, T^r y) \geq 1$. If $\alpha(x,z) \geq 1$ and $\alpha(z,y) \geq 1$, then $x \preceq z$ and $z \preceq y$, and by transitivity, we have, $x \preceq y$. It follows that T is a triangular r-α admissible mapping. Since there exists $x_0 \in X$ such that $x_0 \preceq T^r x_0$, it follows that $\alpha(x_0, T^r x_0) \geq 1$. If $\alpha(x,y) \geq 1$, then $x \preceq y$. Now from the contractive definition of the theorem it follows that $\alpha(x,y) \geq 1$ implies $\rho(T^r x, T^r y) \leq \psi(M_\rho(x,y))$. Hence all the conditions of Theorem D.1 are satisfied, and T has a r-fixed point.

Theorem D.7 1

Let $(X, \rho, \preceq)$ be a partially ordered 0-complete partial metric space and $T : X \mapsto X$ be an r-increasing mapping such that $\rho(T^r x, T^r y) \leq \psi(M_\rho(x,y))$ for all $x, y \in X$ with $x \preceq y$. Suppose that the following assertions hold:

(a) there exists $x_0 \in X$ such that $x_0 \preceq T^r x_0$

(b) if $\{x_n\}$ is a sequence in X such that $x_n \preceq x_{n+1}$ for all $n \in \mathbb{N}$ and $x_n \to x$ as $n \to \infty$, then $x_n \preceq x$ for all $n \in \mathbb{N}$

Then, T has a r-fixed point

Proof of Theorem D.7 1

Define $\alpha : X^2 \mapsto (-\infty, \infty)$ as in the proof of the previous theorem. Assume $\alpha(x_n, x_{n+1}) \geq 1$ for all $n \in \mathbb{N}$ such that $x_n \to x$ as $n \to \infty$, then, $x_n \preceq x_{n+1}$ for all $n \in \mathbb{N}$. Hence by (b) of the theorem we get $x_n \preceq x$ for all $n \in \mathbb{N}$ and so $\alpha(x_n, x) \geq 1$ for all $n \in \mathbb{N}$. Proceeding as in the proof of the previous theorem, we can show that T is a modified-modified weak α-ψ higher-order contraction, an r-α-admissible mapping, and there exists $x_0 \in X$ such that $\alpha(x_0, T^r x_0) \geq 1$. Hence all the conditions of Theorem D.2 hold, and T has a r-fixed point.

Theorem D.8 1

Let (X, ρ) be a 0-complete partial metric space endowed with a graph G and $T : X \mapsto X$ be a modified weak ψ-graphic higher-order contraction. Suppose the following assertions hold:

(a) there exists $x_0 \in X$ such that $(x_0, T^r x_0) \in E(G)$ for any $r \in \mathbb{N}$

(b) T is r-G-continuous on (X, p^s)

(c) $(x,z) \in E(G)$ and $(z,y) \in E(G)$ imply $(x,y) \in E(G)$ for all $x,y,z \in X$, that is, $E(G)$ is a quasi-order [J. Jachymski, "The contraction principle for mappings on a metric space with a graph," Proceedings of the American Mathematical Society, vol. 136, no. 4, pp. 1359–1373, 2008]

Then T has an r-fixed point

Proof of Theorem D.8 1

Define $\alpha : X \times X \mapsto (-\infty, \infty)$ by

$$\alpha(x,y) = \begin{cases} 1 & \text{if } x,y \in E(G) \\ 0 & \text{otherwise} \end{cases}$$

At first we prove that T is a triangular r-α admissible mapping. Let $\alpha(x,y) \geq 1$, then $(x,y) \in E(G)$. As T is a modified weak ψ-graphic higher-order contraction, we have $(T^r x, T^r y) \in E(G)$, that is, $\alpha(T^r x, T^r y) \geq 1$. If $\alpha(x,z) \geq 1$ and $\alpha(z,y) \geq 1$, then $(x,z) \in E(G)$ and $(z,y) \in E(G)$, so from (c) of the theorem, we have, $(x,y) \in E(G)$, that is, $\alpha(x,y) \geq 1$. It follows that T is triangular r-α admissible. If T is r-G continuous, then we have, $\lim_{n \to \infty} x_n = x$, and $\alpha(x_n, x_{n+1}) \geq 1$ for all $n \in \mathbb{N}$ imply $T^r x_n \to T^r x$ for any $r \in \mathbb{N}$, that is, T is r-α-continuous on (X, p^s). From (a) of theorem, there exists $x_0 \in X$ such that $(x_0, T^r x_0) \in E(G)$ for any $r \in \mathbb{N}$, that is, $\alpha(x_0, T^r x_0) \geq 1$. If $\alpha(x,y) \geq 1$, then $(x,y) \in E(G)$. Since T is a modified weak ψ-graphic higher-order contraction, it follows that $\alpha(x,y) \geq 1$ implies $\rho(T^r x, T^r y) \leq \psi(M_\rho(x,y))$. Hence all the conditions of Theorem D.1 are satisfied and T has an r-fixed point

Remark D.9 1

If we replace (c) of the previous theorem with "G is a connected graph", then it still holds

Theorem D.10 1

Let (X, ρ) be a 0-complete partial metric space endowed with a graph G, and T be a modified weak ψ-graphic higher-order contraction. Suppose the following assertions hold:

(a) there exists $x_0 \in X$ such that $(x_0, T^r x_0) \in E(G)$

(b) if $\{x_n\}$ is a sequence in X such that $(x_n, x_{n+1}) \in E(G)$ for all $n \in \mathbb{N}$ and $\lim_{n \to \infty} x_n = x$, then, $(x_n, x) \in E(G)$ for all $n \in \mathbb{N}$

Then T has an r-fixed point

Proof of Theorem D.10 1

Define $\alpha : X \times X \mapsto (-\infty, \infty)$ as in the proof of Theorem D.8. Condition (b) of the theorem implies that $E(G)$ is a quasi-order [J. Jachymski, "The contraction principle for mappings on a metric space with a graph," Proceedings of the American Mathematical Society, vol. 136, no. 4, pp. 1359–1373, 2008]. Let $x_n \to x$ as $n \to \infty$ and $\alpha(x_n, x_{n+1}) \geq 1$ for all $n \in \mathbb{N}$, then, $\{x_n\}$ is a sequence in X such that $(x_n, x_{n+1}) \in E(G)$ for all $n \in \mathbb{N}$ and $\lim_{n \to \infty} x_n = x$. So by (b) of the theorem we have $(x_n, x) \in E(G)$ for all $n \in \mathbb{N}$, that is, $\alpha(x_n, x) \geq 1$. All other conditions of Theorem D.2 follow similarly as in the proof of Theorem D.8 and consequently T has an r-fixed point.

> **Theorem D.11 1**
>
> Let $(X, \rho, \preceq)$ be a partially ordered 0-complete partial metric space endowed with a graph G and T be an ordered modified weak ψ-graphic higher-order contraction mapping. Suppose the following assertions hold:
>
> (a) there exists $x_0 \in X$ such that $(x_0, T^r x_0) \in E(G)$ with $x_0 \preceq T^r x_0$
>
> (b) either T is r-G-continuous in (X, p^s) and $(x, z) \in E(G)$ and $(z, y) \in E(G)$ imply $(x, y) \in E(G)$ or;
>
> (c) if $\{x_n\}$ is a sequence in X such that $(x_n, x_{n+1}) \in E(G)$ with $x_n \preceq x_{n+1}$ for all $n \in \mathbb{N} \cup \{0\}$ and $\lim_{n \to \infty} x_n = x$, then $(x_n, x) \in E(G)$ with $x_n \preceq x$ for all $n \in \mathbb{N} \cup \{0\}$
>
> Then T has an r-fixed point

> **Proof of Theorem D.11 1**
>
> Define $\alpha : X^2 \mapsto (-\infty, \infty)$ by
>
> $$\alpha(x, y) = \begin{cases} 1 & \text{if } x, y \in E(G) \text{ with } x \preceq y \\ 0 & \text{otherwise} \end{cases}$$
>
> At first we prove that T is a triangular r-α admissible mapping. Let $\alpha(x, y) \geq 1$, then $(x, y) \in E(G)$ with $x \preceq y$. As T is an ordered modified weak ψ-graphic higher-order contraction mapping, we have $(T^r x, T^r y) \in E(G)$, where $T^r x \preceq T^r y$, that is, $\alpha(T^r x, T^r y) \geq 1$. Also, let $\alpha(x, z) \geq 1$ and $\alpha(z, y) \geq 1$, then $(x, z) \in E(G)$ with $x \preceq y$ and $(z, y) \in E(G)$ with $z \preceq y$. So from (b) of theorem, we have $(x, y) \in E(G)$. Also $x \preceq z$ and $z \preceq y$ implies $x \preceq y$. Hence, $\alpha(x, y) \geq 1$. Thus, T is triangular r-α-admissible. Let T be r-G-continuous on (X, ρ^s), then $\lim_{n \to \infty} x_n = x$, $(x_n, x_{n+1}) \in E(G)$ for all $n \in \mathbb{N}$ imply $\lim_{n \to \infty} T^r x_n = T^r x$, that is, $\lim_{n \to \infty} x_n = x$, $\alpha(x_n, x_{n+1}) \geq 1$ for all $n \in \mathbb{N}$ imply $\lim_{n \to \infty} T^r x_n = T^r x$, which implies that T is r-α continuous on (X, ρ^s). From (a) of the theorem, there exists $x_0 \in X$ such that $(x_0, T^r x_0) \in E(G)$, that is, $\alpha(x_0, T^r x_0) \geq 1$. Let $\alpha(x, y) \geq 1$, then $(x, y) \in E(G)$ with $x \preceq y$. Now since T is an ordered modified weak ψ-graphic higher-order contraction, it follows that $\alpha(x, y) \geq 1$ imply $\rho(T^r x, T^r y) \leq \psi(M_\rho(x, y))$. Hence all the conditions of Theorem D.1 (or Theorem D.2) hold, and T has an r-fixed point.

Finally, we have the following which applies Theorem D.10 to the existence of a solution of an integral equation

> **Example D.11 1**
>
> Let $X = C([0,T], \mathbb{R})$, and let $d : X \times X \mapsto \mathbb{R}^+$ be defined by $d(x,y) = \|x - y\|_\infty$ for all $x, y \in X$. Then (X, d) is a complete metric space. Assume (X, d) is endowed with a graph G. Consider, the integral equation
>
> $$x(t) = p(t) + \int_0^T S(t,s) f(s, x(s)) ds$$
>
> and let $F : X \mapsto X$, for any $r \in \mathbb{N}$, be defined by,
>
> $$F^r(x)(t) = p(t) + \int_0^T S(t,s) f(s, x(s)) ds$$
>
> Assume the following
>
> (a) $f : [0,T] \times \mathbb{R} \mapsto \mathbb{R}$ is continuous
>
> (b) $p : [0,T] \mapsto \mathbb{R}$ is continuous
>
> (c) $S : [0,T] \times \mathbb{R} \mapsto [0, \infty)$ is continuous
>
> (d) there exists an upper semi-continuous from the right nondecreasing function $\psi : \mathbb{R}^+ \mapsto \mathbb{R}^+$ with $\psi(t) < t$ for all $t > 0$ such that for all $s \in [0,T]$
>
>   (i) for all $x, y \in X$, $(x,y) \in E(G)$ implies $(F^r(x), F^r(y)) \in E(G)$
>
>   (ii) for all $x, y \in X$, $(x,y) \in E(G)$ implies $0 \leq f(s, x(s)) - f(s, y(s)) \leq \psi(\max\{|x(s)-y(s)|, |x(s)-F(x(s))|, |y(s)-F(y(s))|, |x(s)-F(y(s))|, |y(s)-F(x(s))|\})$
>
> (f) there exists $x_0 \in X$ such that $(x_0, F^r x_0) \in E(G)$
>
> (g) if $\{x_n\}$ is a sequence in X such that $(x_n, x_{n+1}) \in E(G)$ for all $n \in \mathbb{N}$ and $\lim_{n \to \infty} x_n = x$, then $(x_n, x) \in E(G)$ for all $n \in \mathbb{N}$
>
> (h) $\sup_{t \in [0,T]} \int_0^T S(t,s) ds \leq 1$
>
> Under the assumption (a)-(h), the integral equation,
>
> $$x(t) = p(t) + \int_0^T S(t,s) f(s, x(s)) ds$$
>
> has a solution in, $X = C([0,T], \mathbb{R})$. Consider the mapping $F : X \mapsto X$, for any $r \in \mathbb{N}$, defined by,
>
> $$F^r(x)(t) = p(t) + \int_0^T S(t,s) f(s, x(s)) ds$$
>
> Let $(x,y) \in E(G)$, then from (d), we deduce
>
> $$\begin{aligned} |F^r(x)(t) - F^r(y)(t)| &= \left| \int_0^T S(t,s) [f(s, x(s)) - f(s, y(s))] ds \right| \\ &\leq \int_0^T S(t,s) |f(s, x(s)) - f(s, y(s))| ds \\ &\leq \psi(\max\{|x(s) - y(s)|, |x(s) - F(x(s))|, |y(s) - F(y(s))|, \\ &\quad |x(s) - F(y(s))|, |y(s) - F(x(s))|\}) \\ &\leq \psi(\max\{\|x(s) - y(s)\|, \|x(s) - F(x(s))\|, \|y(s) - F(y(s))\|, \\ &\quad \|x(s) - F(y(s))\|, \|y(s) - F(x(s))\|\}) \end{aligned}$$

Example D.11 Continued 1

Thus,

$$\|F^r x - F^r y\|_\infty \leq \psi(\max\{\|x(s) - y(s)\|, \|x(s) - F(x(s))\|, \|y(s) - F(y(s))\|,$$
$$\|x(s) - F(y(s))\|, \|y(s) - F(x(s))\|\})$$

That is $(x, y) \in E(G)$ implies

$$\|F^r x - F^r y\|_\infty \leq \psi(\max\{\|x(s) - y(s)\|_\infty, \|x(s) - F(x(s))\|_\infty, \|y(s) - F(y(s))\|_\infty,$$
$$\|x(s) - F(y(s))\|_\infty, \|y(s) - F(x(s))\|_\infty\})$$

Thus all the hypothesis of Theorem D.10 are satisfied, and hence the mapping F has a r-fixed point, that is, a solution in $X = C([0, T], \mathbb{R})$ of the integral equation

$$x(t) = p(t) + \int_0^T S(t, s) f(s, x(s)) ds$$

4.4 Exercises

Exercise D.1 1

Taking inspiration from [B. Samet, C. Vetro, and P. Vetro, "Fixed point theorems for $\alpha - \psi$-contractive type mappings," Nonlinear Analysis: Theory and Methods, vol. 75, no. 4, pp. 2154–2165, 2012], prove the following: Let (X, d) be a complete metric space and T be an r-α-admissible mapping [Definition D.14]. Assume that $\alpha(x, y) d(T^r x, T^r y) \leq \psi(d(x, y))$, where $\psi \in \Psi$[Notation D.15]. Also suppose the following assertions hold:

(a) there exists $x_0 \in X$ such that $\alpha(x_0, T^r x_0) \geq 1$

(b) either T is r-continuous or for any sequence $\{x_n\}$ in X with $\alpha(x_n, x_{n+1}) \geq 1$ for all $n \in \mathbb{N} \cup \{0\}$ and $\lim_{n \to \infty} x_n = x$, one has $\alpha(x_n, x) \geq 1$ for all $n \in \mathbb{N} \cup \{0\}$

Then T has an r-fixed point

Exercise D.2 1

Taking inspiration from [P. Salimi, A. Latif, and N. Hussain, "Modified $\alpha - \psi$-contractive mappings with applications," Fixed Point Theory and Applications, vol. 2013, article 151, 2013], prove the following: Let (X, d) be a complete metric space and T be an r-α admissible mapping with respect to η [Definition D.16]. Assume that $x, y \in X$, $\alpha(x, y) \geq \eta(x, y)$ implies $d(T^r x, T^r y) \leq \psi(M_\rho(x, y))$, where $\psi \in \Psi$, and

$$M_\rho(x, y) = \max\{\rho(x, y), \rho(x, Tx), \rho(x, Ty), \rho(y, Tx), \rho(y, Ty)\}$$

Also suppose the following assertions hold:

(a) there exists $x_0 \in X$ such that $\alpha(x_0, T^r x_0) \geq \eta(x_0, T^r x_0)$

(b) either T is r-continuous or for any sequence $\{x_n\}$ in X with $\alpha(x_n, x_{n+1}) \geq \eta(x_n, x_{n+1})$ for all $n \in \mathbb{N} \cup \{0\}$ and $\lim_{n \to \infty} x_n = x$, one has $\alpha(x_n, x) \geq \eta(x_n, x)$ for all $n \in \mathbb{N} \cup \{0\}$

Then T has an r-fixed point

> **Exercise D.3 1**
>
> Taking inspiration from [A. C. M. Ran and M. C. B. Reurings, "A fixed point theorem in partially ordered sets and some applications to matrix equations," Proceedings of the American Mathematical Society, vol.132, no. 5, pp. 1435–1443, 2003], prove the following: Let $(X, d, \preceq)$ be a partially ordered complete metric space, and $T : X \mapsto X$ be a r-continuous r-increasing mapping such that $x_0 \preceq T^r x_0$ for some $x_0 \in X$ and any $r \in \mathbb{N}$. Assume that $d(T^r x, T^r y) \leq M\lambda^r d(x, y)$ for all $x, y \in X$ with $x \preceq y$, where $\lambda \in [0, 1)$ and $M \geq 1$ is given by Proposition 4.1 [Jeffery Ezearn, Higher-order Lipschitz Mappings, Fixed Point Theory and Applications (2015) 2015:88]. Then, T has an r-fixed point.

> **Exercise D.4 1**
>
> Taking inspiration from [J. J. Nieto and R. Rodrıguez-Lopez, "Contractive mapping theorems in partially ordered sets and applications to ordinary differential equations," Order, vol. 22, no. 3, pp. 223–239, 2005], prove the following: Let $(X, d, \preceq)$ be a partially ordered complete metric space and $T : X \mapsto X$ be an r-increasing mapping such that $d(T^r x, T^r y) \leq M\lambda^r d(x, y)$ for all $x, y \in X$ with $x \preceq y$, where $\lambda \in [0, 1)$ and $M \geq 1$ is given by Proposition 4.1 [Jeffery Ezearn, Higher-order Lipschitz Mappings, Fixed Point Theory and Applications (2015) 2015:88]. Suppose the following assertions hold:
>
> (a) there exists $x_0 \in X$ such that $x_0 \preceq T^r x_0$ for any $r \in \mathbb{N}$
>
> (b) if $\{x_n\}$ is a sequence in X such that $x_n \preceq x_{n+1}$ for all $n \in \mathbb{N}$ and $x_n \to x$ as $n \to \infty$, then $x_n \preceq x$ for all $n \in \mathbb{N}$
>
> Then T has an r-fixed point

> **Exercise D.5 1**
>
> Prove that (X, ρ^s) is a metric space, where ρ^s is given by Remark D.11

4.5 References

(1) Ampadu, Clement (2016).lulu.com, Characterization Theorems Inspired by the Hardy-Rogers Map I: Some Results in Metric Spaces. ISBN: 1365101185, 9781365101182

(2) Matthews, SG: Partial metric topology. Ann. N.Y. Acad. Sci. 728, 183-197 (1994)

(3) P. Salimi, A. Latif, and N. Hussain, "Modified $\alpha-\psi$-contractive mappings with applications," Fixed Point Theory and Applications, vol. 2013, article 151, 2013

(4) M.Abbas and T.Nazir, "Common fixed point of a power graphic contraction pair in partial metric spaces endowed with a graph," Fixed Point Theory and Applications, vol. 2013, article 20, 2013

(5) J. Jachymski, "The contraction principle for mappings on a metric space with a graph," Proceedings of the American Mathematical Society, vol. 136, no. 4, pp. 1359–1373, 2008;

(6) I. Beg, A. R. Butt, and S. Radojevic, "The contraction principle for set valued mappings on a metric space with a graph," Computers and Mathematics with Applications, vol. 60, no. 5, pp. 1214–1219, 2010;

(7) F. Bojor, "Fixed point theorems for Reich type contractions on metric spaces with a graph," Nonlinear Analysis: Theory, Methods and Applications, vol. 75, no. 9, pp. 3895–3901, 2012

(8) Ampadu, Clement (2016):Ψ-Higher-Order Contractions and Some Common r-Fixed Point Theorems in 0-Complete Partial Metric Spaces. Unpublished

(9) B. Samet, C. Vetro, and P. Vetro, "Fixed point theorems for $\alpha - \psi$- contractive type mappings," Nonlinear Analysis: Theory and Methods, vol. 75, no. 4, pp. 2154–2165, 2012

(10) A. C. M. Ran and M. C. B. Reurings, "A fixed point theorem in partially ordered sets and some applications to matrix equations," Proceedings of the American Mathematical Society, vol.132, no. 5, pp. 1435–1443, 2003

(11) Jeffery Ezearn, Higher-order Lipschitz Mappings, Fixed Point Theory and Applications (2015) 2015:88

(12) J. J. Nieto and R. Rodrıguez-Lopez, "Contractive mapping theorems in partially ordered sets and applications to ordinary differential equations," Order, vol. 22, no. 3, pp. 223–239, 2005

www.ingramcontent.com/pod-product-compliance
Lightning Source LLC
Chambersburg PA
CBHW051103180526
45172CB00002B/756